Experiencing oneself as a silent, motionless unit of God, the individual can maintain the dynamic capability of full motion and activity. One finds that the restlessness which is so much a part of human nature is instantly and permanently reduced. No longer needing to hold to a fixed value of time as the basis of activity, the human being whose heart is fully evolved is free to develop a meeting of time through the true psychology of the Infinite. In this nonmoving, almost breathless state, the individual develops a lock on time and can live permanently in this clockless, infinite value. One views everything as continuous, all present, ever-altering, yet completely changeless and perfect. This experience of an expanded yet static universe defies description, yet it is the heart of the heart.

—*Timeshift*, Chapter Two: "The Divine Heart"

TIMESHIFT

The Experience of Dimensional Change

Janet Iris Sussman

TIME PORTAL Publications, Inc.
P.O. Box 2002
Fairfield, IA 52556
United States of America

Printed in the United States of America
by Thomson-Shore, Inc.

ISBN 0-9643535-0-4

CIP 95-06002

Cover and book design: Laurie Douglas
Illustrations: Susan Hoover
Executive Editor: Douglas A. Mackey

To those who seek the Divine
in their actions, internally spinning Time
towards its matrical beginnings.
May the journey to uniform dimensionality
open us to the mysterious
and subtle regions of the Heart.

CONTENTS

Chapter Two: The Psychology of Global Reception

Chapter Three: The Complementary Body

Chapter Four: The Geomorphic Energy Shift

ACKNOWLEDGMENTS

The writing of this book has been supported and developed through the loving encouragement of many people. The original version of the book is now really a separate document. Not a single line of that text actually remains in this new book. However, without the one-pointed, relaxed focus of the original three people who sat through the most raw perception, this second version would not be possible. I therefore wish to thank Lilli Botchis, John Lentz, and Robert Wilkinson for their patience with and belief in the manifestation of the life of dreams.

The writing of this new version, though a solo process, was unknowingly birth-mothered through the encouragement of Dorothy Certain, who rented a very special house to me where I could feel safe and comfortable. I also wish to thank Robin and Marty Skinner for their technical support in the world of computers.

The first round of copy editing was done by my very capable mother, Selma Sussman, incorporating initial suggestions from John Halberstadt. The editorial team of Dirk Haueter and Doug Mackey, which formed Time Portal Publications, was invaluable. Further editorial assistance was quickly and expertly furnished by Katharine Hanna. I particularly want to thank Doug for his many probing questions that challenged my ability to explain and assimilate this work, preparing me for presentation in a wider context. I also want to thank Dirk for his steady, loving presence in helping me to retain the essence of this material, which, in the end, cannot be put into words.

The visual content of this book is the result of the combined efforts of three very talented people – Susan Hoover, who brought the chapters to life with her startling illustrations, Dale Engelbert, whose preliminary design work gave the project structure and coherence, and Laurie Douglas who moved the work forward to its present, distinctive conclusion.

I wish to especially thank Lilli for her knowledge of plant

medicine that nurtures and sustains me, and for her constant push to break the boundaries of my mind and heart. I also wish to acknowledge her for whisking me off on writing vacation/ adventures that helped me throughout this project. Special thanks to John Lentz for his insightful introduction. Thanks go also to my many other friends, who guided and encouraged me all along the way.

Lastly, I wish to thank Dr. Richard Wertime of Douglass College, Rutgers, The State University of New Jersey, my college creative writing teacher and advisor, whose prophetic vision that we "train the unconscious" has served me well throughout my life.

—Janet Iris Sussman

PREFACE

Six years after having witnessed the initial creative process that produced *Timeshift*, I am pleased to give the reader a glimpse into how this book came about and what it is like to participate in the consciousness it generates.

Timeshift is the fully crafted version of an earlier book Janet Sussman wrote in 1989–90. Though the material had been living inside her for many years, this earlier book was the first time she had tried to put it into a form that others could experience. She asked me and two others to support this effort by sitting with her while she translated this knowledge into language. Our role was primarily to hold stable the consciousness field around Janet while she wrote the book. I hope that what I have learned through both witnessing the creation of the preliminary version of *Timeshift* and through experiencing this published version will help the reader access the consciousness out of which they have arisen.

We met for almost a year. While Janet sat at the computer, the three "witness/participants" looked over her shoulder. During each session my mind had to confront its many limitations. Part of me resisted the whole event because the new language was asking me to adjust how my brain processed words and sentences. I was never able to grasp what it all meant from a purely intellectual point of view because the units of quantum intelligence coming into my nervous system were larger than what I was used to handling. Yes, the words and sentences were familiar, discrete units themselves, but they were part of a larger matrix of intelligence that could not really be unraveled into language the way we were doing it through our brains and through the computer. Sometimes sentences that on the computer screen were flowing out row upon row actually never lost their extended value. Instead, they seemed to stretch out into the environment beyond the mental and mechanical apparatuses through which I was experiencing them. My mind would leap after a "thought," thinking it could collect what in actuality was a thread of intelligence stretched out to infinity. It was hard for me to grasp that this mental expansion was actually more important than corralling the material into intellectual formulation. It is this kind of experience that reading *Timeshift* can naturally produce.

There are conceptual ways of understanding this phenomenon, but I have had to accept that such understandings are not part of the experience itself. Understanding takes place in a different and, we might say, lower domain of consciousness. The paradox of the issue is this: the tools by which I am able to understand anything are themselves created in the higher domains I am seeking to know. In the light of this insoluble perplexity, what follows are a few concepts that help me. They are not "truth," for the experience itself transcends the lower mind's ability to understand.

Human consciousness is a very complex, multi-dimensional event. Esoteric teachings in both the East and West reveal a sophisticated layering of subtle body vehicles that carry certain vibrations which allow "reality" to be processed in numerous ways. On the densest level, our physical body reveals to us the material world in all its opaqueness and diversity. One step subtler than the physical body is the vital or etheric body, which contains within it the blueprint of intelligence upon which the physical body draws to process the energies needed for it to function. The astral body communicates the emotional values of experience, both positive and negative. The mental body processes basic thinking functions, but in its higher levels it also acts as a translator of the cosmic intelligence into thought and feeling, which then provide the basis for action in the physical world. There are many more planes of reality beyond the higher mental regions, but the planes discussed here are closest to our experience of *Timeshift*.

It will be clear after only a few minutes reading *Timeshift* that emotion plays no role in this experience. The book's vibration is quite a bit higher than what can resonate in the denser matter of the astral body. Therefore, looking for a familiar "feeling" value will be futile. We can also say that the role this material plays in the lower regions of the mental plane of our consciousness is at best a tenuous necessity. The words on the page and the concepts involved are necessary "handholds" (or perhaps "mindholds") for the consciousness to grab onto while the more dynamic vibrations of the higher mental plane are energized. The biggest challenge for us as readers is to realize that the higher mental plane processes do not involve language as we understand it. In fact, the higher mental plane functions as a creative reservoir in which all potentialities of form are collected in mathematical and geometric constructs. These constructs are of such a complex, multi-dimensional nature that we cannot appreciate how they could possibly be the building blocks of the reality in which we so naively live our personal and social identities. Yet they are. The origins of *Timeshift* are these sorts of constructs, incommunicable in three-dimensional reality and only expressed in this book as a shadow of a shadow of what really exists on the plane of its origin.

So, how does Janet Sussman create this material? Using this model of subtle planes, we could start by saying that she is awake in those spaces of consciousness where cosmic energies take form and are prepared to be sent down into the lower vibrational fields of the mind, emotions, and body. Instead of picking up knowledge in its lower, diminished energy forms, where it is distorted by the misunderstandings of an unknowing world, Janet picks it up in its original energetic construct, much closer to its place of origin. It would be an egregious oversimplification to say that the material spontaneously presents itself to her from some "outside" source as she writes. She has been forming this knowledge through a creative "interaction" between her higher and lower minds for over ten years and continues to integrate that experience daily. Therefore, when she writes, she is actually translating a higher part of her own consciousness into a form more familiar to consensus reality thinking. Her particular challenge as a spiritual being is to relate these higher regions of her mind to the conscious self through which she lives her daily life. What integrates these two disparate experiences for her is yet another value of inner life, pure consciousness, which is the ground state of individual being and is the foundation of all experience, regardless of the state of consciousness out of which one is operating. This level of awareness is a highly advanced condition of the individual spirit and is cultivated through regular and sustained practice of meditation. When the process by which Janet opens up her consciousness to its higher mental expressions is extended into the reader of *Timeshift*, the reader's own awareness moves into his or her own higher mental functions. This movement is enhanced dramatically if the reader has a high value of pure consciousness through which that movement can be integrated into daily mental processes.

To read this book, therefore, is to participate in the forward movement of human evolution by adding our own awareness to the collective higher mental awareness of the planetary consciousness. With that infusion we give "weight" to the higher mental plane construct for a more "enlightened" world, a weight which can then "incarnate" if a sufficient threshold of energy is reached. In other words, through this book, our minds can infuse vital life into the thought forms that will recreate the Earth into a "New Age" or "Heaven on Earth."

This concept of recreating Earth through higher mental functions implies a self-referral loop within the total field of human consciousness. If there is any concept essential to dealing with the challenges of *Timeshift*, it is this loop; namely, all life is a construct in consciousness which seeks to transform itself through endlessly relating back to its field of origin, which is pure consciousness itself. If that process is impeded, the forms and patterns of living cease to renew themselves, and the possibilities of life become severely constricted.

Timeshift is about many things, but mostly it is about human consciousness renewing itself by altering the temporal structures of our reality. As a construct of reality, this book is time in its unbounded value. The unbounded value of time is also the book's underlying literary theme. In the first chapter, "Time-Space Dimensionality," time is existence wanting to be everything possible; time allows life to flow; time activates potentiality in all things. In its "uniform dimensionality," that is, when it constantly renews itself through its own self-referral nature, time unfolds the endless possibilities of creation. Each moment should be forever renewed and revitalized in the pure potentiality of time. When time cannot renew itself, each succeeding moment is bound to the past in a sequence of events that becomes increasingly narrow in its potentiality for the "future." Consciousness, however, can recreate time in its own likeness and release us from the binding sequentiation of "past" events. Through this process of "desequentiating" our "past," we can create a platform of infinite potential by which "future" events can unfold. Through time life can lift itself into the light of eternity. Whatever impedances we might experience while reading *Timeshift* occur because of a conflict between time bound within the narrow constructs of our personal reality and time in its unbounded value as it manifests in the act of reading it.

Though time as a main literary theme slips quietly into the background of the other three chapters, the processes described and initiated in those chapters actually unfold by recalibration of our temporal codes. Thematically, though, here is a short summary.

In Chapter Two, "The Psychology of Global Reception," the infinite value of reality becomes manifest through acts of perception involving both the mind and heart, creating a link between the individual and the unbounded values manifest through both of these channels of experience. When perception is no longer governed by the unprocessed experiences of the past, the heart can open itself to the subtle "feeling" values that are the very essence of perception itself. Through such unfoldment the individual becomes linked to a uniform field of life, of which he or she is but one point value.

Chapter Three, "The Complementary Body," unfolds an astonishing vision of both the human and terrestrial bodies as self-generating, self-transforming and linked into the light/energy system of the whole cosmos. When the human body is able to recognize its own infinite intelligence value and through that recognition totally join with the Earth, the human race and the Earth will become a cosmic traveler no longer limited by the laws of nature that govern space, time, and matter as we know them.

Chapter Four, "The Geomorphic Energy Shift," directs the reader's attention to the Earth as an evolutionary being about to undergo a profound transition into a higher dimensional state. Earth is presently a dimly conscious entity within the galactic matrix that seeks to link itself into the larger field of evolutionary intelligence of which it is a part. Human consciousness is vital in its role as part of the Earth matrix undergoing transformation. Only highly evolved levels of human consciousness can participate in the planetary transition because the laws of nature that will govern life after the transition will be dramatically different.

Each of these chapters has its own thematic focus, but the underlying value for the reader lies in the transcendental functions through which these profound transitions will take place. Reading, understanding, and processing cosmic streams of intelligence are linked together within a single experiential field, the field of one's own consciousness. Though the themes themselves direct the cosmic streams of intelligence to specific areas of our own life processes as we read, the overall influence is one of awakening to one's own self, one's own being, one's own truth.

From here on, the reader is best left to his or her own experience. I have shared mine here as a kind of "leg up" into the saddle to go along with Janet's own assistance in her introduction.

Finally, this version of *Timeshift* is, as I have already said, different from the original work I participated in. The question about this material has always been how much should Janet do to help the reader understand intellectually and how much should she encourage the reader to surrender to the experience as an energetic event in consciousness. Eventually, she decided that it was necessary to wrap an extra fold of meaning around the unbounded field of pure consciousness that is the book's core. That effort was a four-year project. As a result, *Timeshift* has a more sharply focused intellectual lens than the original version had and is, therefore, more organized and more intellectually accessible. The original version came in rather wild and unformed. This one has the virtue of having its time value re-cognized within its own "uniform dimensionality." *Timeshift* is a unique contribution to the field of consciousness studies, not only through what it says, but primarily through what it is.

—*John Lentz*

FOREWORD
How to Read This Book

Read this book as you would a poem. Let the words create a feeling value in your mind, but it would be a mistake to try to understand the concepts from a purely intellectual level. In order to be appreciated, the book must be allowed to work on you. Enter into communion with it.

It is not that you must believe everything written here in order to get something from it. In fact, belief has little to do with it. Do you have to believe in a game of tennis, in a symphonic concert, in a swim at the beach? These are experiences that must be taken in their whole value. Let the reading of this book be that type of experience, one which creates a memory of pure cognition, a feeling of receptivity. Find a tempo which suits you and settle into the pages. Understand that in doing so, you will be affected physically, emotionally, and spiritually.

So take your time. Put the book down and try to absorb it, then take it out again. Let it be a therapeutic process. The words have been crafted so that the translation can retain the flavor of primary experience as much as possible. Remember, though, that the primary experience is one that in our time occurred over many years. Therefore, be gentle with yourself and enjoy.

INTRODUCTION

As we look over the Earth, we question our role here and the opportunity arises to examine the nature of consciousness and its place in human evolution. This expanded essay is an attempt to open a window of inquiry into the nature of human perception and our place in its future. Earth at the present time is full of possibility, danger and wonder. My experience over the past fifteen years has led me to believe that the Earth is a living, conscient being that is seeking to communicate. This project is an attempt to extend to my readers the dialogue I have been having via "night school" visitations.

Like any study situation, the process of receiving information and creating a practical use for it is both empirical and subjective. The voice that imbues this manuscript filters through my waking and sleeping consciousness. The impulse-generated responses that create this experience have led me to delve into worlds where I normally might not go. When these sessions occur during my sleep periods, they are experienced through dream circuitry that is experiential, body-centered and lucid. My mind merges with the information and I can see, hear, even smell the information as it coasts through consciousness. It can be likened to going down a water slide, the thrill of which causes me to forget to hold onto the side once the experience has begun.

I have engaged in a lifelong process of refining my ability to hear, interpret and translate. I see myself as a bridge between humanity and intelligent vortices of vast dimensions that are trying to help us understand who we are and where we can go. This process opens the recipient to higher levels of human cognition. It is my Self that is listening. I am the one who focuses on a given object or situation. I am also that which must clear the way in order for experiences of an extra-generative creative nature to occur.

The process of creating this manuscript began in 1989, precipitated by a visit from Hurricane Hugo in Charlotte, North

Carolina. For ten years I had been trying to describe my understanding of the Earth and our changing perception to friends and associates, as well as people attending the spiritual counseling groups that I facilitate. Somehow, when Hugo struck, my psychological self was prompted into action. Originally, in an attempt to ground this creative process, three other people around me attempted to create a meditative atmosphere as I sat at the computer. The original material poured out so swiftly and intensively that it was difficult for me to maintain the stability in consciousness necessary for me to do the work. The present version was accomplished more easily and without outside assistance.

As a spiritual counselor who utilizes a state of heightened or altered awareness in the course of my work, I have a degree of proficiency in entering a state of merged reality with another person. This has served me well in the present undertaking. The difference, however, is that this project has required the stretching of inner boundaries to develop a new vocabulary, a new language structure, in order to express concepts that are difficult if not impossible to describe in our present system of communication. The translation from a visually holographic language to a flatter plane has been the greatest challenge.

My purpose is to stretch consciousness so that the ability to wrap around cognition is expanded for both reader and writer. To create an ordered value out of relatively difficult or new concepts demands that one elicit an inner response that is both mindful and not of the mind. In this paradoxical place between the worlds lies a type of poetic or lucid grandeur in which I believe the senses can open themselves to greater dimensions or ranges of feeling. The material to be absorbed must lie inside the brain awhile and be kept from analysis. This requires a leap of trust into the realm of inner knowing, inner seeing. As this is accomplished, understanding is gained from a place of internal empowerment which causes words to sprout meaning in whole new areas.

It is not my intention that the material be validated or agreed with through normal pathways of reason. It is better to lift oneself out of the sense mode and be open to the possibility that the material can restructure consciousness through its pure sound value. If the process is proceeding smoothly, it feels like entering a musical chamber in which the eye/ear hits the palate and a melody of cognition begins. As this occurs, the person receiving begins to feel, smell and taste the value of the experience, much like a therapeutic dive into a cross-weave of brain mechanics. The flow of experience feels like being in an airplane as the cabin pressure shifts. The sense of being remade in a way that is both physical and translogical, is the vantage point that can be achieved. This is by its very nature an experimental activity.

This book creates a living experience of the shift in time values as they relate to human cognition. The seed of awakened knowledge will yield something different in everyone. The book seeks to create a vehicle for the actual shift in physiology that will bring about a biological renaissance. Although war rages on the Earth and many suffer from disease, instability or hardship, we are being asked to lay the groundwork for a truly positive and astounding future. We each have a unique responsibility in this process. The challenge here is to inspire the journey to frontiers of knowledge in which we move from the static to the infinite. We are given the opportunity to develop highly original plans of activity based on our own inner sourcing to make new memories for planet Earth.

The activity of cognition is both personal and collective. It demands that we reorganize the time intervals so that as we devour the present, we can view the past and future more clearly. To find the outreaches of personal significance we must learn to be comfortable with perpetual change. Our strength lies in fashioning a vision that modifies itself through adaptability rather than fear. If the trends of collective experience can be cognized while meeting the need for personal identity, our success is assured.

One cannot transcend the personal without first experiencing it fully; we cannot lose sight of our purely human nature. This book, however, challenges our notion of what it is to be human and makes available to us new forms through which interior meaning can be recodified. As we jump out of our skin, so to speak, and recalibrate our perspective, many new vistas become available. This leap into omnipotence involves a kind of creative derangement that will ultimately lead to order. It is not insanity, but a type of restored sanity, a resting place for a mind troubled by its own self-consciousness. As this wave of meaning/thought expands, we can cure the ills of our inner voice and become value-free even as we become more sure of ourselves and those around us.

I feel that love in a broad sense plays an important part in this. Yet love, as a clear field, is not easily attained. The stripping away that must occur in order to lay new groundwork in time can "cure" our hearts, preparing a foundation for pure feeling. We have only to develop the desire to reap love and it can hold us gently in its grasp. Love is the encounter with the sacred object, the "joyself" that is present in all of us. It is a free commodity, an oceanic visitor of perception, so one cannot look for it. Once it is sought it escapes; once it is caged it looks to be free. Love must be met through a range of experience with the "bliss value," the internal happiness that is gained through merging with the Absolute. Through this path, love can open the gate towards a *globobiotic* union with interstellar consciousness. The inherent God value is in nature-prime, the reliving of our aliveness through implied form. Love reidentifies itself; we can remake our activities, our aspirations, our planes of difference. This is our mandate, but also our purest pleasure at this moment.

Many people with whom I have shared this journey are looking for answers right now. There is much confusion as well as a sense that we must go inward to reveal true direction. In writing this book, I am attempting to help people link to new experiences of space-time. For millennia, sages have viewed this world as pure illusion. I have come to view it as a completely subjective and perhaps necessary reality bridge in which our perception has for too long been locked. In this opportune juncture of space-time, the lock on the heart can be lifted free and the cogno-feeling value can spill forth. Then we will develop new territories of life perception and escape from the moorings of the mind.

This process, though sometimes exceedingly difficult, is very rewarding, even ecstatic. In all of its spirals of growth, it asks us to be flexible, tolerant and willing to modify our aims when they no longer concur with altered perception. We must adapt to our changing experience of time in order to more accurately embrace the unifying qualities of love. This is the beauty afforded us during the transition.

CHAPTER ONE

SPACE-TIME DIMENSIONALITY

CHAPTER ONE

SPACE-TIME DIMENSIONALITY

Organization of Time and Space

Our present construct of time is a product of the synthesis of human imagination and creative perception. Time, as we know it, is not an independent function; rather it exists as the by-product of those who witness it. Time is referenced through shared experience; collections of memory and patterns of activity provide familiar landmarks. This type of "clock time" is culturally based, providing a framework for human accomplishment.

Our present view of time is limited because we are lacking the pure experience of a mathematical fabric for time. This is a felt experience of mathematics, a grasp of time through the ground mechanics of consciousness. We might call this ingestive mathematics and it is completely experiential. Without this ability we are unable to move forwards or backwards; we are caught in an unchangeable, fixed version of what we perceive to be reality.

Actual time is unbounded and without object. It stretches infinitely across the range of intelligent experience, swinging itself through numerous channels of perfected structure. Actual time creates a realm in which we perceive a hint or flavor of timeliness. This is a realm in which we recognize the given sets of possibility.

Actual time is inductive rather than deductive. It is circular in that it can loop back over the wheel of its own beginnings, allowing us to stretch a hammock over the void. Since it is nonreferential, it is blind to itself. It does not harbor prejudices because it lives on its own objective wave of cognition. Able to ground each option freely, actual time does not enclose the user, but extends itself into an infinite value, allowing the laws of nature to speak freely and interchangeably for themselves.

The circular, ever-bending subswing of time allows us to lift free from caged possibility. Every individual or collective entity can live a series of events without having to loop back over them in order to establish personal identity. This creates a "moveable feast" in which time can overlap on itself, allowing parallel dimensionality to develop.

We can learn to map time through our unified expression, creating identity from our experience of Self separate from the field of events. Uniform dimensionality is evoked, an experience in which time is unified through a central core or identity mass. This nonreferential field of expression with response to time creates *breakfronts* through which time loops off from one field, only to reestablish itself at the next qualifying interval.

Though this may be perceived as random, this type of time is highly structured. The unit of structuring is *pure consciousness* itself. Time chooses its mates from the interstices of pure awareness, the gaps or breaths that are interlaced into the field of pure cognition. The individual or collective establishes itself in a multidimensional framework. Breakfronts are located, enjoyed, but passed over rather than held to by the movement of the mind. When time is perceived as a wraparound value of cognition, it can release itself into the field of momentum that invokes constant change. This new model of time creates a healthy backdrop for psychological, linguistic, and intellectual realization to grow.

Time is a random messenger. Time appears organized, with historical markers or reference points at every glance; however, these markers are essentially falsified passports to the relative world. Referential time, with its nomadic collection of psychological and historical markers, creates a collection of seemingly related material strung together in an analytical union.

Time, left to its natural system of ordering, strings together rows of attractable pearls, conduits of energy manifested through vast telegraphic networks of sound, sense and perception. In this multidimensional unified reality, the user can climb aboard any row of strings and develop knowledge of the future or the past. Indeed, there is no past, because we experience everything in the "all now," embracing reality, never stopping for contemplation or analysis.

When the user can become one with this movement of time across the stars, he/she gains the capability to let go of time capsules created through limitation. We are free to open into the realm in which all creation is recognized, both manifest and unmanifest. These sparklers of time, these bits of time/matter/energy, are the stuff out of which all form may be developed. Time becomes the clay of the infinite.

In our society, with the advent of advanced communications systems, we perceive data to be organized into interlacing fields that are called networks or matrices. These webs of interlocking identity have within them an organizing structure which is usually thematic and is based on a linguistic or mathematical integument. Simply put, a matrix is a system of data which has been shaped to express a qualifying perception by the user. We might define this type of matrix as fixed. A non-fixed or *extended matrix* would involve a pool of data which exists independently in scattered or random sequences, but which comes together through its own referencing as need or demand allows.

Actual time spins itself out of the thread of extended matrical interface. Matrices are composed of "dots" of space-time building blocks, which in turn compose the mapping material from which quadrants of time can be coded. The networks that are constructed form vast stretches of material, collected in such a way that the read-out can be gained from any point of view.

The matrix rests at the absolute value. It is produced through the natural ebb and flow of on/off reception, representing the something or nothing that glues the universe together or causes it to come apart. Time forms matrices in order to speak to itself, to develop options in the trend of advanced perception. Time is colorful in this respect in that it can portray the range of its prismatic field all at once or in vast exotic arrays of space-time objectivity.

The purpose of human evolution may be viewed as the process of lining up with the *time codes*, of emptying the collective imagination so as to cause it to drop off from the field of memory. It is not that memory by its very nature is unattractive, it is simply that in its present form memory has become hypnotic. It produces a locking away of necessary options through mind-numbing repetition.

The true function of memory is to hold the preconditions of infinity, rearranging them for random use as experience is fed the product of its own time codes. Memory in this capacity is anything but dull. It dazzles the mind with its own storehouse of ready data and can instantly transfigure the value of reality. Memory becomes reality because it reproduces itself spontaneously on the surface of desire and leaps into the unknown continuously. This type of memory produces intelligence that is greatly needed now.

Matter as a Timed Event

Matter is not only composed of energy, it is also composed of time. Matter is a timed event. As matter is sped up or played down in the balance of the moment, the events of its own origin are superimposed upon it and played back in its internal memory. Matter maintains an energetic bank account that makes a statement every moment of its development.

Time frames its material existence by creating *time shifts* or mobile units of space/time/motion which are redirected from one category of space-time delivery to another. Time "wobbles" as it learns to select variables from different compartments in the space-time continuum. This slide of time, this shift of the time variables from one interface to another, allows time to live value-free, independent from the entire scheme. Time becomes free to choose its own random influences from the bank account of pure time or pure matter. Time and matter maintain a joint account.

Time is concrete in this sense, at least as concrete as space or motion. It is only viewed as relative because its *host factor*, which is matter, can be seen more readily from our field of dimensionality. Time is concrete only in the sense that it is the building block of human existence. Without a sense of time, our entire ability to experience reality would be impaired. We would never have time for anything because we would be constantly waiting for an event to happen.

Time runs the distance because it is capable of clocking every moment of its own value in the panoply of life's existence without questioning why. When time values are uniform, there is a constant sense of speeding up without any sense of uncomfortableness or motion sickness. We have the ability to shift dimensions, to timeshift, to literally swing ourselves from one dimension to the other.

Our lack of understanding about how to move through the subswing of time limits us greatly. Since we are unable to absorb full spatial dimensionality into our awareness, we cannot make the leap into fully conscious life. As we experience time as concrete, space and motion are able to sandwich themselves into the fabric of waking existence. The mind is jarred into a different style of electromagnetic functioning and is propelled into a sense of holographic unity.

There could be no time shift if matter were static. The shift is based on the possibility that matter can undergo a direct time-pull at its base and shift over into the qualifying norms. Matter will not appear the same because it will no longer be strung out on the window of time. Matter will become playful. It will be able to create itself into constant and, at the same time, ever-changing values of form and scope.

Essentially, matter is only static at the "end zones" of our space-time dimensionality, yet it appears completely static to us because of our present time perception. This is true because we are not able to absorb the time values in their proper sequences. If we were to do so, we would increase our propulsion through time and allow for synchronicity in the interplay of change/

motion. The parallel motion of time dances on the wheel of life, creating interesting paradoxes in the space-time variables. The shift in time involves our ability to see how matter lapses or destroys itself, only to be built up again in the wake of time's reemergence. Matter disappears; it enters into discrete phases of time alteration until it can be recast in a different makeup or sequence of materialization. These lapses in matter occur randomly and are filled in by our own minds. We create the solidness in matter in order to maintain a sense of psychological equilibrium.

As we witness time collapsing back on itself, retrieving its uniform values, we see that it picks them up again on the down side. This gives us the unique opportunity to view time as a function of itself. The time sequences loop over the back of their own structure, interweaving sequentiality to produce pockets of matter/time which we perceive as reality. Through this process, we interface with our own experience, developing symbology which enables us to interpret areas of personal and planetary development.

Interlocking Time and Space

When we begin to create uniform spheres of experience and every unit of time is relational, the *loop-locking* interface allows us to perceive time as going backwards and forwards at the same rate. Time and space become equidistant cousins to each other. Then we can travel back through time without losing anything. Time is never lost and events are desequentiated, depending on their rapid movement in the field. When we enter the parallel time field, we are essentially free to travel as we wish. There is no proportionality in this unique interface. Everything is boundless, free of motion, unlimited in scope.

In other words, in the wraparound interface, the value of an event is not desequentiated in relationship to the causality of the event. The event supersedes itself, it self-refers to its own balance of motion. It becomes supralinear and is free-floating in its own field.

In short, time develops through the interplay of motion, motion develops through the field of change, and change develops through the expression of distance. Distance creates the pockets of matter/time which repeat themselves in infinite cycles of variance.

Stored Values of Perception

Human beings have become dependent on developing tribal memories that identify a limited range of activities as proper to human functioning. The entire range of knowledge of what is truly human has been lost. We human beings are barely able to call up the range of feelings that constitute our birthright. Our range of feeling has simply become programmable from the options we have given ourselves in what we have constructed as the past. In this way, the subtle blend of thinking/feeling/enunciation which can lead to original expression has become unknown or rare at best. The synthesis of highly curved memory, which is fountain-like, effervescent, life-inducing, and filled with creative possibility, lies dormant in our mental field.

As time is erased from our identity, at least time as we have known it, we are capable of entering the time/perception of field-generated activity. The focus of such activity is to call up open sets of time, recodify them and feed them back to our central nervous system. This type of inner voyaging into the perceptive masks of previous time makes our language capable of releasing the synchronous energies that induce parallel matrices of mind-activity directly to our physiology. The mind becomes a prismatic reference point for itself. We reference stored syntactical memories, but rather than simply releasing them back into preconcerted patterns, the patterns are broken up, creating breakfronts of perception that create new word interlays. These linguistic networks, which are the psychological equivalent of time itself, relay time back to the mental forefront, so that it can spin out new variables and produce entire systems of visual and auditory experience. Indeed, we induce senses that heretofore would seem nonsensical.

This version of reality, which is matrical, unified and clearly cognizant of itself, even as it is ever-changing, interdimensional and free-spirited in its essence, produces a field of constant play. It is filled with joy because it never ceases to be amazed at itself. It is innocence without measure. When time is nonreferential it is reverential. It is holy because it is filled with holes. It counters its gaps with material reality but always enters this filling-up with emphasis on leaving the holes, the void spaces, intact. Thus, there is always room to come up with more.

Crisscrossing Mind Fields

Time can directly relay empirical data to itself. Thus, the mind that knows time knows itself. Time can reduce itself to preconceived variables even as it gobbles them up. This sense of time devouring itself appears in classical notions of primordial mythology. It is the dragon swallowing its tail.

Time, portrayed as the fire-breathing monster, relives its magical synthesis of stored reality breakup through the element of fire. Fire is the changing variable in the scheme of cosmic reality. This is because fire is the relaying energy of transformational activity. When time portrays the mind, it portrays it as a matrix of crisscrossing patterns of rejuvenative reality. This type of mind restores itself at every turn, but it must also burn up its dross at the same time. The dragon that burns the flame of metamorphosis spits out the "fires of time" so that it can succeed in relating its story through metaphorical parallel. Yet metaphor itself is only a linguistic symbiosis for the transformation. Time itself is the mechanism for transformative follow-through. Time allows parallels because it must create stripes and squares, curlicues and dots, waves and particles, to recreate its dragon skin, its smooth yet textured surface

When we say that time equals distance and distance equals the span of motion across time, we mean that time is non-linear, yet can invite activity through its synthesis of mind/matter nerve endings. When time crisscrosses back on itself, it re-remembers itself and thereby creates consciousness. This type of consciousness is highly organized yet completely random at the same time. All things happening at once: this is the layered activity of a healthy mind or a healthy universe.

Time Codes

To create a code you must have a blueprint. Time always creates its own blueprints because it knows itself. It is as if you had a map available to you for every interval of breath that you might take, for every step you might walk.

Time knows itself, but it is lively with this knowing. The joke is on us. Time relives itself only because we want it to. It has no memory other than what we provide for it. Therefore, when time reunifies itself through the balance of its own *mathematical symbiosis*, it restores its time codes or response variables directly to its own core matrix or electromagnetic resonance chamber.

Time is the strict balancer of its own score sheets. It has no place for mistakes because it is self-correcting. Time restores its mathematical symbiosis through interlocking blueprints of codified material, similar sheets stretched out over a wide continuum. These sheets, or bandwidths, enhance our present notion of clock time to that of a matrix-field-spread, causing perception to be interlaid with inductive possibility.

Time is quite reasonable. It knows the possibilities at hand, yet it seeks to surprise itself by living its own proof. The glossary of time contains the intercedent matrices of open-ended reality. In this sense, reason can be relied upon only as a nutcracker, a tool to get the contents open. In its place, reason is reasonable; it behaves itself. However, when time allows reason to manifest itself as the central relay, chaos can result. Nothing destroys orderliness so rapidly as reason. This is because reason insists on destroying the natural order of things, events, people, etc. Therefore, time must constantly restore, overlook, reunify. Even at the present time, in this state of disorderliness, time is restoring itself, repairing itself, seeking to put the holes back in so that space can permit a change.

We view matter as something to fill in the spaces; time views matter as something to revivify the space. Space is the principal component, matter secondary. Matter reduces itself to fit the intergalactic race for space in the message units we call civilization.

Planet Earth is a time-coded message unit that is constantly signaling its past/present/future matrix to other civilizations. The signals we relay now are by-products of that which we have relayed before. When time is recrystallized at its very essence, it will allow the planet to recognize itself for the genius that it is. The signals we send will not be based on past correspondence but will be transduced through the innovative mechanism of a nondistilled reference point for time. Time will restore order through disengaging limited orderliness. This is what we have to look forward to.

The Timeless Traveler

When the individual or collective leaves time behind and enters into the now, the ever-present nonchanging, nonrelative reality, time ceases to be simply pervasive, it becomes omnipresent. Time becomes the dominant skin through which the texture of experience can be viewed. When time is open, it can breathe its own creation through porous intervention with others.

The timeless traveler, not limited by his/her counterparts in past dramas, paints a picture with ever-brilliant colors which can be seen for many "miles" in the continuum. The effects of the timeless traveler are everlasting because they are not limited by one mapwork in the diameter of space. They are field-resonant, open to encounter, and therefore vastly reachable by others.

The present matrices of time being played out on our planet are not easily viewed by future/past/present; our activities are limited by our own lack of vision. We seem trapped in an isolation booth, waiting to be discovered by other civilizations, other dimensions. This is because we have limited our field of resonance to that which can resonate with us and, for the most part, this is simply ourselves. We are always reviewing, rarely renewing. Until we can escape this isolation pod and enter into a matrically-induced array of time-change reunion, we are likely to die here, our range of perception closed and unwaverable.

Death comes only to those who have not understood time. Once mastered, time gives way instantly to immortality. There is no death, but only shape-shifting, motionless reality, placed on the "skin" of space. The craft is in the crafting. Time is the stuff from which our own personal infinity can unfold, but it takes a certain level of fearlessness, of mind-courage, to accomplish this.

Unleashing Hope

In a civilization in which hope has flickered, time can quickly catch up and restore adventure. Time must do this, however, through erasing doubt. Doubt occurs when the range of possibility is lost. Doubt closes off the range of motion and thus produces a "time-bog" in which hope cannot enter.

When time is reperceived as hopeful, as living, as valuable, while being free from fixed value, hope springs forth naturally. Everyone can be hopeful when it is understood that there is always more time. When the person or collective views time as ending or limited, anxiety is the instant by-product.

Hope cannot be manufactured through relative emotion. Hope must be a mathematical by-product of possibility, magnified by knowledge. When hope conjures up its own causality, it is very easy to come by. Hope is in fact startling; it presses upon any perception of limitation and squeegees it out, leaving the fabric of creation behind.

Our present civilization appears to lack this type of hope, but it can be retrieved simply through entering into contemplation of field-time within its own value. Any restoration of order certainly depends on this.

Indicative Trends in the Field of Time

In our present prepatterned history, we interpret the value of time through points of identification with our storehouse of events. This is how we weigh the possibilities and anticipate certain outcomes from them. This response to time, which is nonregenerative, is not enticing to new forms of activity. There is no novelty to it.

However, when time is self-indicating through encounters with its own fields of activity, it can become very sparkling in its ability to seek out other trends and lead the way to them. Time can actually seek out its counterparts on the expressway of happening and recall or reconstitute them at will. This clamoring for the trends of time, this seeking out of possible parallels, makes time remarkably cozy at remaking its numbers. Time can seek out its trends through mathematical symbiosis with its attractive counterparts. Rather than going forward and backward, time remains at the center, at the groundpost, the "point-zero break" between continuum and infinitude. Time, from this midway vantage point, looks out over its armies and calls up the troops by enlisting its markers to dance order into the cosmos.

When time can be perceived as setting a trend, not embryonically, but at its birth/inception matchlight, then civilization will be on to something. Time sets trends as one would set a fire; a simple blaze can grow very rapidly when fanned with a subtle wind. Time looks at trend-setting, views noted possibilities, and fans the fires when the need arises. In the present state of things, time can only fan the fires of that which historical interface has identified as "trendy." Time cannot manufacture new trends easily because the space from which to do it is not readily available. There is no space in the spacebank, no time in the matrix with which to interweave. So time is resting, remembering, rebuilding itself for the interval when civilization can restore itself through proper maintenance of its time codes.

Time bends back on itself through the power of infinite flexibility, and in this respect it is its own causality. This is fortunate, because without such flexibility, time would literally have snapped a good while ago. When a bow is not given any space to stretch, its tendency is simply to snap in two. All of the processes which are now in place to reunify time are giving it some room into which to stretch when the possibility of order is ripe. Time will pay attention and will *queue up* its origination on schedule.

Translating Referential Perception

To move from a static point of reference to one that is freely moving, the referential boundaries must be broken. When time becomes capable of speeding up its return to an infinite value, the referential codes are shaken up in order to proceed to the proper point of correction. The new time codes may change rapidly as the point of correction is achieved.

As time expands, the rate of flow-change also increases. We may think of this like the flow of liquid from a glass. Once the flow increases with the acuteness of the angle, the liquid pours more steadily and freely. So it is with time. As the angle of change is precipitated, the human flow of evolution is caused to speed up or increase as well. This may cause startling changes in the landscape of human events.

To incur changes with the minimum amount of stress, there needs to be a release of tension from the matrix without fully restoring all of the proper values. The release of tension, signified by a "strange turn of events," can awaken collective perception and cause a shift in the field. Time refers back to itself. It establishes itself uniformly along the matrix of change and waits for the proper period in which to restore order to the landscape. Time refers to points of order within the referential landscape as boundary points for this change.

With the speed-up in the time codes, people are beginning to look for answers within the sociological stream of events. However, time looks for answers within the matrix of organized values. It seeks out stragglers and tries to rectify them. Time corrects itself from the sea of referential value points and restores order. This develops a proper chain of command for new and more evolutionary matrices to form.

Time as a Rainbow

The view from the window of perception is colorful. The colors correspond to the interior majesty of the shades of perception as they are registered by human vision. During a time change, perception is altered. Time does not appear to have boundaries, but instead becomes diffuse. The landscape becomes translucent rather than opaque. With the advent of change in perception, time itself becomes more physical rather than less. In other words, we can see time through the window of matter, because matter is not perceived as perfectly solid. We can see the bands of time, the stretches in the fabric of reality.

This colorful attribute of time, its ability to reveal shades of meaning from within itself, allows historians of our day to interpret events, to cast meaning upon them. This is how relative time is presently cast. With the shakeup in time values, time will become even more colorful and the significance of referential points will become more obvious.

What may be misunderstood, however, is that one views time itself, not the event. Time can become limited by the window of perception if the perceiver is focused only on the event and not the flow of time within it. One must become capable of separating time from the field of events and view it as distinct. Perception can then be free to alter the interior reality and develop a more balanced perspective.

Ingesting the Mathematics of Time

Time may be experienced as a mathematical symbiosis of energy/matter transition. Time locks upon the spiral of activity defining the shape, size and color of objects. Time then exhibits an ingestive mathematical structure which exhibits its cognitive value as an expanded blueprint. When mathematical expression is taken in through a subtle level of feeling, it can be ingested directly into the nervous system and experienced as a felt realization. In this experience, intellectual and *psychosolar* mechanisms are unified.

Time cannot create mathematical unity without developing a system of matrical entities that outline the pure consciousness it wishes to express. The mathematics of time is simple. Time records the unit of measure as a spiral or gate through which the unit of time can pass. For example, in present clock time, each minute or hour is expressed through a particular unit. Nonreferential time uses space itself to "carve up" its own units. Time identifies fields in space and delineates them through shades of perception or vibration. Time identifies space through lifting the light on its true character.

Space is not as empty as we might think. Emptiness is not the character of space any more than it is that of the mind. Space is full of charged radiant activity. It is really a super-charged world in which tiny particles are intercepted at a rapid pace and bounced back and forth across each other's bow. Time utilizes the activity of space to scan the screen of adventure and lock onto tiny quadrants in order to develop matrices or breakfronts of consciousness/matter.

Time creates matter out of space through relative expressions of distance. One light year, for example, which is the distance it takes light to travel in one solar year, might be expressed as one small vista in the field of energy known as time. Time depends on developing its own matrical shields which allow it to speed up or slow down at will. Time uses space as a vehicle for time travel. It stores "ordered space," shakes it up, and spills it back out again at other reaches of the galaxy.

The mathematics of time is concerned with the rate of change of time values in relationship to the quantification of space. Time reflects the value of space back onto itself and restores order in the process. It is an analytical guessing game of vast proportions. Time receives the valuation of space from its own derivatives. In other words, it analyzes the character of space, describes it through mathematical correlation, and develops time codes which store the memory of this information for future use. Without these time codes, time would simply forget its own nature; it would return to the chaos from which it sprang.

Clock time was developed to infer the unity of space-time and thereby lay the groundwork for human or intelligent organization of any kind. All manner of civilization has developed through the advent of such timely organization. It becomes possible to understand how consciousness can travel through different dimensions by understanding time as a series of dot-dash cues—binary codes of information which are fed on-off values through the inter-relay of space-time dimensionality. Time relays the value of space to itself and this information is relayed to a central intelligence system which translates it and feeds it out into the space-time vault. There it is recodified, stored and redeveloped for use in the *message units* of future civilizations.

Time masters the concept of space by identifying where space is, by underscoring the relative curve of space. Time infuses itself directly into the fabric of space, thereby creating matter. Matter infuses itself back into the space-time value, thereby creating room for consciousness. This interrelationship between time and space magnifies the need for change. It creates the opportunity for change in the relationship between Man/Woman and Nature.

Alternate Fields in the Space-Time Void

Although we perceive time as something stationary and fixed, time really has the character of motion. Time occurs alternately between light and dark fields of motion within its own vantage point. When time peeks out into the field of relative activity, it develops a window through which it can see these dark or light fields and stir things up by advancing a field of communication between them.

The dark field signifies the spectra of relationships between the unknown universe and the known universe. The light field represents the quanta of activities within the known universe itself. The known universe encompasses the spectra of activity generated through the advent of advanced civilizations which have begun to quantify this information and to publish it. To us, the known universe is what the New Continent might have been to the ancient explorer. It is the territory of our own imagination, but it is also the territory we have mapped and codified through scientific effort.

Time skips back and forth between the known and unknown universes and thereby develops alternate spheres of influence which appear to us as future time, past time, present time. These sectors of activity interact with each other and make up the boundaries we presently perceive as reality. As the known universe expands, so does our experience of the future. We can then experience time as something lively, energized by the value of our own imagination, but also recalibrated through the space-time sphere in such a manner that it can be developed for future use.

Time is always useful. It relays information back to itself for use in future vectors of activity. When time calls itself up, remembering its own identity, it reinvents a means of recording itself for future use. Without this inventive nature in the field of time, matter would have nothing on which to attach itself. It would not be able to identify its own structure. With the development of the light and dark aspects of time, matter can create the on and off movements that signify creation.

Out of the darkness comes the light. Time reconstitutes itself through this off-and-on movement, developing chains of command which can reunify the entire continuum. Matter identifies itself and begins to spring to life.

Understanding Units of Mass

As human evolution leaps into the twenty-first century, the types of matter formations which are generated will depend on humanity's grasp of time. If time can be perceived as the superfluid, lively and entertaining entity that it is, then humanity will be capable of living outside the boundaries of time with the necessary mechanics engaged for time to be a useful friend. However, if humanity seeks to cage time through the use of technology which limits its expansion, there will be a type of darkness, a type of pent-in feeling that may last for many units of space-time value.

In order to create a more perfect evolutionary setup, time must be allowed to spring back into its full flowering. This may be done through speeding up the *time curves*, allowing matter to become more lively in its internal nature, and by human activity itself.

It is possible for humanity to enter the twenty-first century in a state of unification between energy and matter—there need not be a separation. When humanity can perceive itself as reunified with its central component, the free-floating, ever-present character of time, then the entire race will leap forward into new avenues of activity. There will be ceaseless wonder.

The Bite Between Time and Matter

The interface between time and matter may be viewed as a path through the avenues of creation. The Supreme Creator created matter so that it could be stretched and developed for many purposes. When matter is confined to time, it cannot be stretched; it becomes brittle and breaks. Time therefore, must be kept open and free of impairments so that matter can expand to create new parameters in the doorways of perception.

When time is allowed to do this, matter becomes friendly, isotopic; it justifies itself by changing its position. It creates compounds, magnifies its intensity and changes the character of unbending objects. Time in its relationship to matter is unending, it can change shape while retaining its essential character. When time is opened up, matter is freed as well. In this way, new creations, new events, new life forms can occur.

Life is connected intimately with its own field of time. When time is allowed its own gestation, new life can begin. It can go from the protozoan to the complex in a matter of hours. It does not need any superhuman time frame for such accomplishment. New forms of life, energy and composition can spring up at any time, simply by nourishing time from its own Mother, the wellspring of consciousness.

When time is capable of restoring matter to its original form, which is lively and distinctive, any manner of creation can occur. Humanity will become its own creator, redeveloping "time wheels" and allowing them to spin for the good of all.

Educating Matter

Educating matter is a learned response with respect to time. Time becomes capable of losing its motion as matter is created. Thus, matter has to be able to trip up time with relation to its motor responses, suspending time so that space for matter can be created. One may say, therefore, that matter is educated through the field of time.

Matter learns its response to time through time codes that relay information to the field of matter and determine the characteristic boundaries that will be implied. We may think of this as a playing field in which matter circulates across time and establishes rudimentary boundaries in order to establish its goals. As each set of goals is reached, matter determines its appropriate time characteristic from the set of uniform values that are spread across its surface. Matter scores time and relays its response variables directly to the stored time vaults for future use. Since matter is the recording secretary of time, it establishes its own limits. It allows time to score its goals through intercepting matter as matter reaches its established end.

Since time is infinite and matter is not, matter must constantly remind time of its appointments so as to not be late in the field of scheduled matrical interlinks. Time is the unit of measure for the scoring of matter in the universe. Time prepares itself for reaching its goal of interceding matter, developing a friendly relationship, and preparing the ground for matter to come back to itself to become time once again. We may say that matter and time are interchangeable.

If matter and time are to become united and form a matter/time chamber from which advanced civilizations can develop their own principal time-feeds, matter must develop chains of command that circulate time directly through the interstellar spaces of universal flow. The relationship of time and matter may be compared to two friendly nations, exchanging goods and services, but maintaining their individual sovereignty. The difference is that time and matter can perceive their mutual identity; they rely on this foreknowledge to keep from becoming one with each other too quickly and becoming undifferentiated.

It is this differentiation, the separateness of life, that defines matter and gives it breadth of relationship in response to time. Without this separate identity, life would simply be timeless, undifferentiated and devoid of matter entirely. Creation would not have taken place.

Life Seeds

As time creates matter, so does matter create energy. As energy is created, the seeds of life are generated and allowed to sprout in the bed of time. Matter creates *life seeds* in order to spawn the creative matrices that signal the advent of civilizations.

Each civilization, whether it is human, animal, or a species unknown to us, has within it a developmental blueprint that is formulated through many strata of genetic implanting and reseeding. Each species is shaped through the developmental influence of time/matter. A species is developed because it is timely. The species is simply a point on the watch of time. It signals a new formulation for the arcs of time to travel through.

Time likes to have company. When it gets lonely and seeks companionship, it develops lifeforms which are forms of time enacted in the field of matter. The goal of time is to reprint itself over and over again through new forms of life, painting a picture book of its changing perception through history. Time records itself so that its spread of influence will be more widely felt. Rather than staying in its pure but invisible form, time makes itself visible through the pattern of creation.

Time is the co-creator of its own symphony of matter, producing the tones that will be most pleasing to the ear of the Creator. Time signals the proper tone that will make music in the ears of matter; then matter reproduces itself so that time can become available to it once more. Matter is like a restless infant sitting in the crib of time; it cries out for the possibility of creation and enjoys itself once it is made manifest. Without the proper webwork, the human component, the guests of time, would not be realized.

What is an Arc?

An *arc* is the pattern through which time can travel. Time needs a map and the arc provides that map. The biblical metaphor of animals entering the field of time "two by two" refers to the ark or arc through which matter can pass to restore time. The arc forms a doorway through perception itself. It provides the tunnel through which time can travel.

Since time is curved and matter is straight, matter can be looped through the tunnel of time. However, the arcs that are formed through this interlay are field-generated response mechanisms for the *slope of time* as it is cast off from the field of matter. We may think of these arcs as enormous jetways from which matter can take off, store cargo and return safely to its home base.

Time appears sloped because it loop-locks with its own internal magnetic *spin*. When time relays its information into its own filing mechanism, it creates a *shield reponse* which allows it to imprint itself with its own memory. This creates a balance point in the union of opposites. When time arranges itself in such a configuration we say it is "sloped" because it appears to configure itself from every available variable. From its omnidirectional vantage point, time can "call out its ends," revealing its arc.

As this approach begins to occur, time becomes "cross centripetal" and capable of revealing itself through the diameters of motion as well as of internal *impedal velocity*. The relationship of time to its endposts in this predetermined spin creates a play of opposites. The material generated is like the negative space of a painter's canvas. Through this portal the relationship between time and matter is continually reshaped.

As time coasts along, encouraging its drop into the end spin at every point, time creates an internal drive, a push in the direction of entropy. In this way, the spin becomes centripetal even as it thrusts downward in the direction of time/velocity/motion. Each spin of time, each resurgence of time from its endpost is transduced as a slope or uphill climb in the direction of infinity. This is why time can be seen as twisting on its ends, relieving itself of internal pressure and escaping any finite means of grasp.

This infinite range allows time to complete its qualifying intervals at ever more rapid rates. The impedance slows down so as not to compete with the variables that are transduced by this interplay. Time remains even, free of constraint, and capable of carrying its primary medium, motion, without having to superimpose new constraints or prior knowledge. Time becomes radiant, ever-buoyant, and able to coast along, while the slope remains changeable, variable, capable of returning to the "needs of the time."

The message of time becomes one of translucence, internal dynamic spin, and vestigial reemergence. This allows the slope to remain even and entirely parameter-free. Without this quality, time would appear to be opaque, solid, without any translucent interplay between itself and the concurrent avenues of matter that it seeks to represent.

The slope of time becomes the firing mechanism for matter to return to its less opaque, more centered, and evanescent regions. Time enters this slope with full harmony, entropic veracity, and vision. It can see itself even as it creates new dynamic frameworks. Time opens up and recalls its own nature through sliding down the slope of its own internal parameters in the wonderful sleigh of the infinite.

Matter likes to travel through time because it is in the business of manifesting creation; it likes to make things. Time, on the other hand, is constantly trying to restore order; it is the unmaker of reality. Time wants things to return to their unmanifest state; matter, on the other hand, wants to bring things about. Matter is the active state of time.

The arcs act as bridges between the active, field-generated responses of time, and the older, more passive portals, most of which are still in use. The active tunnels are the arcs that are heavily trafficked by the response mechanisms of time, characterized as matter. Since matter flies off at a moment's notice, it cannot wait around for permission from time to restore its boundaries, therefore it hires time out to restore order every so often. The arcs establish points of order in the developmental matrix of space-time relationship. Without the arcs, there would be limitless expanses of space-time and no seeding regions for creation to form.

Each planetary system or star matrix creates boundaries of space-time and within each system of boundaries, arcs form. Planet Earth itself is an arc tucked beneath the banner of time.

The entire planet is a seeding station for matter to be recodified and sent out into this region of the space-time biosphere. The arc of Earth is the place in which matter, known as codified history in our time frame, reconstitutes itself to be fed back into the interlocking units of space-time from which it sprang. The arc of Earth is the message unit of the mother to her space-time children. The Mother is saying, "Let me imprint you with my own version of space-time," and the units of space-time oblige. She is saying "I am life," and the time/matter interface is responding by saying "I am life."

One of the secrets of life is replication. Without replication everything would die. However, replication is not actually produced by the union of opposites. Replication is produced through the union of samenesses, similar fields of space-time magnitude cross-boundaried on the interlude of time. Time produces identical replicas of itself for use in the development of matter. This is why, though very different, each species is potentially the same when it comes to time.

Time knows all of its categories very well because they are all response mechanisms to the one central intelligence link. All differentiation is therefore surprisingly and infinitely similar. This is one of the secrets of the Universe.

The Nature of Nature

We have been talking so far about the development of civilization, as if that civilization did not have to reside within the boundaries of the natural world. Nature is the response mechanism of field-generated matter to itself. Nature is the mother of invention; she is the mother of matter and the foremother of time. Nature creates the rhythms through which the centrifugal clock of time can flow. Nature creates the pulsations that synchronize the value of time.

The time signature of nature is the heartbeat of matter. Matter listens carefully to time, understanding its essential call in the substructure of the Universe. Nature listens carefully to matter, recording its necessities and then inventing them in the form of earth, flowers, trees, etc. The natural order of things is therefore the response of matter to its need for a perfected environment.

The restoration of matter is rooted in the need for Nature to prepare for the intercession of time. Nature ages through the need of time to reconstitute itself into pure form, that is, pure energy or consciousness, once the call to the millennium sounds. Nature provides the waiting arms for the child matter to come to itself, to reveal its essential character.

For this reason, Nature is extremely mutable. She can be bent, almost to the point of breaking, because she is the plaything of matter. She holds the crystalline intelligence upon which matter can record its history. Yet, Nature is indeed breakable. Like all things which are stretchable, Nature has its breaking point. This occurs when the time values of the infinite are no longer able to be recorded due to a series of imperfections that slow matter down and thereby impede the flow of Nature. This is the present state on our planet. Nature is being hindered from reproducing herself because of the difficulty in moving through the obstacles imposed by the intonations of matter. Therefore, in order to break up the patterns and restore order, Nature is calling on time to ask that the essential character of creation be rerouted.

Nature does not interpose herself lightly on the field of matter because it is her job to hold off, waiting for the right interval to set things straight. However, when the entire future of this sector of the galaxy is being affected by the condition of Nature, she has no choice but to step in and restore a balance to things.

Time is the fundamental principle behind the scheme of Nature. When Nature creates a rock, a stem, a tree, she calls up the parallel matrix in the scheme of dimensionality for that creation. Time is held open, waiting for Nature to begin. When time is ready, the response values of matter are interposed and Nature springs herself to life. She creates the blueprint, the seeding mechanism for a new species, an environment that is right for that climate, position, or strength.

Nature has the value of accomplishment in relationship to matter, but Nature cannot hold back the trends of time. When time wishes to express itself, it will simply crack open the field of Nature and pour the contents of its timeliness directly on the field of matter. In these circumstances, the changes will not be slow, but will occur quickly and sometimes without warning. Therefore, Nature is a power to be reckoned with because she speaks the truth as far as time is concerned. This is why you cannot fool Mother Nature.

The Stages of Matter

First Stage

Matter in its most refined form is the inclusionary unit of time. In this form, time is stored in order for certain principles of Nature to be established. In the refined value of matter, time is based on the rhythms or cycles of Nature from its inceptionary level. Nature, being the causality of time, creates the space necessary for matter to be set up and to record itself in the analogues of creation.

Matter in its most refined state is therefore completely unified and undifferentiated, while simultaneously being capable of manifesting itself in its totality. Nature relies upon this unmanifested, free-floating quality of first-stage matter to create the field of possibilities for creation. It may be said that in this stage, matter is holding off time, waiting for a developmental kick to happen.

Second Stage

As matter picks up in its vibrational intensity, it begins to interact at a more rapid perceptional base with the time values. At this interlude, matter is lukewarm in its activity level. This is the range in which normal human thought occurs. At this level, matter is cognized and through this specification is no longer unified. Deprived of absolute value as its primary identification, matter can be split up at second stage and rerouted to secondary time codes for encapsulation. However, matter at this stage is still directly tied to the absolute and may go into refinement at any point.

Matter creates its response variables with respect to time at this secondary stage. It recalls the field of time and imprints this aspect onto the chain of thought/material reality for use in developmental creation. When a human being creates a concept or idea and percolates or synthesizes this idea, he/she is drawing upon this secondary quality of matter. This is the advent of thought becoming a thing.

Third Stage

Once the time codes have been established, matter begins to turn back onto itself. This may be thought of as a spinning motion in which matter recalls its identity, creates a chain or field of command that is responsible for detail, and develops plans of encoding for future events or periods. At this stage matter is visible and tangible; or, in terms of creative process, action has been made manifest. However, being action in a state of infancy, it is not locked into primary time values and can be easily changed or reversed if necessary.

Matter has the capability of remembering itself and therefore all of the variables that might be representative of it. When matter identifies these variables in this third stage, it transfers the range of feeling to a future reality, planting seeds for a fourth generational construct to take place. At this point, although matter is fully lucid, it is not completely congruent. There are avenues of intercedance built into the fabric which can be netted out and released. This lack of congruency is necessary in order not to fix the time values into place.

It is not until the fourth stage that matter is fully lucid. By our standards, third-stage matter is therefore not completely solid as we would define it through our field of vision. Although no stage of matter produces complete solidity, matter reengages its matrix values at select intervals in order to develop patterns of intercedence that lock solidity into place. This is necessary not only so that matter can be visually perceived but also because matter would break up if the change rates were too high. This would cause an involuntary instability at all junction points.

Fourth Stage

At this stage, matter becomes fully opaque in visual terms and is highly regenerational. In other words, since its patterns of formation have been solidified, its coding mechanisms can be duplicated. This sets the stage for what we know as reproduction. Matter reproduces, as has been stated, through the process of evolutionary sameness.

The pattern of opposites normally construed as prerequisite for regeneration are balance codes in the present field matrix of human evolution. This is to say that the present construct, which validates a male-female union to uphold the field constraints, may not always be necessary for reproductive pathways to open. The polarities established in this matrix are now limited because more evolved regenerational pathways of genetic implementation have not been fully established in human individuals. In effect, the range of consciousness does not presently have full cognition of itself.

Fourth-stage matter is characterized by this lack of full self-knowledge with respect to its character. In a sense, it is fully existent, but it does not have the complete map of its existence available. It has yet to be taught the secrets of its own nature. This state of ignorance would seem useless; however, it has been built in as a safeguard in relationship to the full measure of creation. By not understanding its own unity value, the fourth-generational creation cannot attempt to create or restore values of which he/she is still only partially aware. In other words, unable to span the full range of possibilities, the fourth-generational creation is responsible only for the character of desires and not for its response to them.

When the transition to the fifth stage begins to occur, the individual actually begins to have the capability to manifest matter. In this way, the individual becomes unified with the base of matter which is the level from which first-stage matter patterns itself. Traveling through the conduits of presupposed time, the individual develops a sixth sense which matures into a detailed developmental network of understanding as to the field-matrix value of consciousness as it repatterns into activity.

We are presently undergoing a shift from the fourth to fifth stage, i.e. from an orientation based on desire and temperament to one based on unified objectivity. It is the shift from being generated by our whims to being generated by our causality. It is the shift from irresponsibility, born of ignorance, to full responsibility and full knowledge.

Fifth Stage

This stage is the beginning of a time-sequence matrix balance that is based on a unified relationship between time and matter. At this stage, the individual or the created matter is capable of relating directly to the time source and developing interrelational patterns of response with respect to action, field formation, and generational conduct.

Here, the human individual would be said to be enlightened, but this is only the beginning of his/her travels. When the relationship between field-generated time and clock time has been mastered, the individual is capable of restoring the time values when they are thrown off through the motion of activity. Activity's primary job is to shake up the perfectly unified and stable order of field-matrix relationship. It is this changing motion that creates the periodicity of evolution.

As the individual gains access to the sprightly time inherent in this fifth stage movement, he/she can walk between the worlds and establish a basis of operation that is at once material and nonmaterial. In terms of visual perception, this individual views matter as solid only insofar as it need not be cut open through simple visual acuity. At the same time, by changing the lens through which perception is viewed, this opaque, fairly solid relationship to matter can become immediately translucent, unstable and subject to breakup.

In the transition between fourth- and fifth-stage perception, this may appear confusing to the individual. However, once this field of perception is established, he/she can perceive the actual particles of time through which and from which matter regenerates itself. From this point, particle-interlude seeds develop future pathways of space-time interaction. The fifth stage consciousness may appear to be slow-moving in clock time because it does not require a great deal of action to effect directionality. The relatively frenetic pace of the previous stages is based on their having to constantly rebalance that which has been rendered unstable through activity.

The fifth-stage consciousness is interdependent; it restores its unitary value through a direct relationship with the evolutionary flow of Nature. It is inherently right-reasoned with respect to time and therefore cannot interfere with it. However, in the gain back and forth between the fourth and fifth stages, which happens in a natural evolutionary flow, the individual may experience a depth decorrelation with respect to his/her time frames and may appear confused, disoriented, or emotionally unstable. This is because the emotions, as we know them, are restorative values with respect to matter. This will be discussed in full in Chapter Two.

Evolutionary Mind Flow

The mind with respect to time is an encapsulation of itself. In other words, the mind in its pre-fifth-stage value records the character of events through inherent comparison. The mind ebbs and flows, a tidal wave of uncertainty, in which the parameters of consciousness have been slowed down so that the perceptive values can be recorded for future input. The subjective state, however, as we know, is mind that is constantly in motion, without silence or repose. The mind is the pre-evolutionary pool in which the primal time codes or value relationships are archived.

The mind, therefore, is fully capable of synchronization only when it is safe with respect to unification. When the mind, or psychology, withholds unification, it does so because it cannot fully frame unification at its present value. In other words, the mind cannot restore time until it can live within its boundless and infinite formation. If the mind were to restore the infinite and inceptionary nature of time too quickly to itself, it would simply slip away, unable to organize, retrieve, or implement data. This would cause a stage of extreme mental confusion. The mind must assimilate the value of time in small steps; it learns the value of the new field, tries it on in the previously established context, and imprints upon itself a new field of activity formation. The mind experiments with time and develops a more fluid range of motion.

Civilizations are based on the collective will to develop methods of activity, which although they appear separative in nature, drive continually towards unity. This differentiation develops the range of individual psychology and matter/time field derivatives, which then form the escape ladder for consciousness. When consciousness is not fully able to restore its parameters through a full union with itself, it creates a lesser value, which engenders a separative economics of time, fields, and energy.

When the individual no longer experiences this separative nature of time/matter as fulfilling, then the range of unification inherent in the time construct can actually begin to take place. Time organizes itself primarily in response to units of motion by changing the scope of mental activity. This activity slows down so that gaps in the range of motion can be contacted. Consciousness can then recorrelate itself to the slope of time which its own nature has registered.

The mind enters a state of grace with respect to cognition, in which the value of unified cognition is a natural state. As this occurs, the mind becomes thorough, all-knowing and responsible for its correlation to time. In a psychological sense, the individual is then capable of establishing an experience of mind that steps outside event-generated response. This stand-alone quality of mind allows it to jump ahead in response to its own character, developing a range of activity based on timelessness. This is the character of mind we presently seek.

Jump-Starting the Mind

The mind circulates along particular evolutionary pathways due to the baseline psychology of the individual. To keep orderly and restore relative motion as change occurs, the mind is dependent on an even focus with respect to time. In its pre-fifth-stage state of resolution, time is required to jump-start its focus of attention so as to establish itself on the field of prominent events. We respond to these events, and in effect create a time-code interface that characterizes the nature of that particular field of consciousness.

In order for the mind to be established in the field of no-time, it has to agree to release its basis of attention in this approach/avoid range of activity. We have to relinquish the drama of possibilities and enter a state in which the flow-continuum in relation to time variables can become more perfectly realized. To do this, time essentially steps in through a maximized state and brings us the experience of timelessness. This timeless feeling initiates a response that signals to the mind that it is safe to let go of its stasis and enter a path of unity.

In this new state, relative motion will be governed solely through the causality of time variables. As this occurs, the pathways to intelligence can open; the mind becomes capable of locating itself through experience that is both timeless and perfectly timed. Time has new points of reference. Intelligence is then speeded up and can link directly with the primordial stages of mind in first-stage awareness.

Time as Ghost-Buster

By allowing the mind to follow its own direction in time, the shadows of consciousness that were established on the mental field can recede. These shadows are blueprints of activity which have not yet manifested, or which have already manifested but have left electromagnetic residue. These ghosts are detrimental to the free-flowing nature of the mind. They create false pictures of stasis and thus fool the mind so that it cannot restore true unification.

The mind that is time-bound tries to remember that which has previously occurred, whether it is a psychological or an event-centered reality. In the mature fifth stage, the mind loses control, creating an open field for the interflow of time/matter to occur. Once in a more receptive state, the mind is capable of erasing those time values which inhibit the flow of organization. The mind can free itself from the past, establishing a qualifying present; we are able to create a signal system that evokes future/past/present in a more evolutionary way. The mind can release its grip on the past by blocking future pathways until it is capable of fully absorbing the significance of its ancestry.

We cannot live in peace with ourselves or others until we completely regenerate our mental nature. A present which is limited, filled with fear and held to through rigid time/value interlacing cannot seek the future in a coherent manner. Therefore, the entire bracing of field-time is dependent on the point at which the mind enters manifest consciousness.

Exploring the Field of Every-Mind

The mind is a phenomenon that is not limited to human beings. In the field of matter, it may be said that every organism, no matter how primitive we may think it is, has a mind. Mind is the stuff through which time explores the *matrilinear* endpoints of consciousness. Matter creates time values within itself, stipulating possibilities in consciousness. Matter follows through on these junction points through union with the mind. Therefore, every qualifying disposition of matter can be interpreted through the field of matter; every dip into matter invests itself with the quality of mind.

This understanding allows us to realize that matter is not unintelligent in any form; it always presents a codified view of things, organizing itself in such a way that is dispositional to its principal field. When mind infuses itself into the field of matter, it strives towards unification. Every cell, every primitive organism, is moving towards this point of unification. In more advanced fields of intelligence such as the human species, this intelligence-making is done through coalescing certain building blocks of order. The system is stretched, allowing for the development of field-generated complexity in the human organism. Each cell operates in a similar manner.

When intelligence is realized as a time-centered, matrilinear event, it can identify any aspect of the breakfront of Nature.

From this point, time can then be crystallized to represent matter. That is, pure time, or pure nonmatter oriented creation, can be made visible. The perception of matter as non-material, simply energetic, or time-derivative in nature, is the basis of dimensional movement not usually thought of in our sector. As the time values change, we are freed up. This allows us to understand time as the basis for consciousness. It also magnifies our perception of consciousness in species with which we are already familiar, such as domestic animals.

Living the Life of Time

Once time is understood as the current running through all life, then it can be realized directly. This direct link to time affords us the possibility of entering an explosive and startling landscape of color, shade and meaning that can sweep our awareness and free us from blindness. Time is the thread of fire that lifts us out of the continuum of omnipresent darkness and provides the illumination necessary for us to experience freedom.

Since time is the clue to birth at our present stage of perception, it is important that we become friends with the field of time and do not allow indecision to leave us suspended, unable to grasp our orientation. The inevitable shakedown in our frame of reference is intentional and beneficent. We cannot escape from it, nor should we try. By coming to meet our own time machine, which is our perfect mental intelligence, we can enter a world in which time can be traversed much as a trapeze artist swings from one platform to another. By lifting into the arcs of time and allowing planet Earth to be our traveling platform, we can look deeply into the intelligence of the Universe and open ourselves to a link with our destiny that can be truly inspiring.

We cannot leave time behind. Every event, every pattern of occurrence that has interposed itself on our field of awareness is unified within us. However, by stretching ourselves to fit the rotational cortex of time as it stretches itself along hidden boundaries, we can become capable of freedom in its most profound sense. This type of freedom, which is highly creative, innovative, and interdependent, makes time stand on end and allows us to look beyond our limitations to points of discovery that are inherent in our deepest nature.

The Thread of Symbiosis

The interdependence of all life is not simply a figure of speech. Time causes all matter to move as pearls on a thread arranged in spiral. By unifying with time, we cause ourselves to function deep within this movement and our independent style of functioning no longer exists. We enter a world in which every action that we take, every possibility of momentum, causes us to affect that which has come before and that which will come after. Since there is no beginning and no end, we are capable of influencing every thread of existence. If you shake the pearls at the center, the two ends will also jiggle.

When we understand our interconnectedness, time can open us to its rendering value—the creative influence of time on the unmanifest. Probabilities spread themselves throughout our individual and collective universe. Interactions occur that heretofore we did not think possible. When we are no longer attempting to escape time but climb within it, the vaults open and the possible links between our venturesome spirit and our creative intelligence can come into play. When we become playful with time, we are willing to cross the boundaries of perception into a colorful world in which every ray of energy/matter is filled with prismatic contents. We can then render our own colors in relation to time.

As we look at the spiraling networks of interconnection, they are no longer time-limited. They can then be viewed in their process of perfecting their interlocking basis for reality. We are the circuitry through which the matrices of the world implant themselves and perfect union. We are the manifest desires of the infinite.

The Looking-Glass of Perception

We cannot step around time in the way that little ones do, stepping around the mother for fear of stepping on her toes. The time values demand risk, they demand synthesis and they demand unification. They are very demanding. They ask us to open our hearts to the possibility of matricing with forces or fields of matter with which we have not felt capable of linking, nor even knew existed.

When time relinquishes its secrets, the past/present continuum melts, leaving us to view ourselves inside out without any points of reference, without possibility of escape. When the mind can view its own referential boundaries and can peek into the looking glass of its essentially mirror-clean nature, we can look below its shiny surface. The mind is capable of reducing itself to its own image only when it is perfectly clean with respect to referential boundaries.

When these boundaries are broken open through the demands of time, the consciousness can establish itself independently of any past experience. In this shape it is capable of changing shape, as shall be explained in Chapter Three.

Cleaning the Mind

When we attempt to clean the mind, we are asking that all of the matrilinear links that have been established at the surface of awareness be gradually eliminated. We stand before the mind, humbly, without expectation, and demand that the time values enter to restore order and continuum to consciousness. When we superimpose our belief systems as to how to do this, we immediately risk the threat of destroying the plan. Therefore, the mind must be cleaned through its own internal mechanisms. It cannot be accomplished through introducing anything that is contrary to its own nature. All formulas or dogmas must be eliminated in order for the mind to become responsible to time in a direct sense. Time enters the law of union when it is allowed to peek into the recesses of consciousness and clear the necessary parameters. By eliminating the possibility of doubt, the mind restores its own clarity.

This type of cleaning could be established at approximately the seventh year of life. Until this point, the human individual does not know itself as a fully independent entity. It is still an extension of the mother and therefore has not had the time to develop a fully differentiated consciousness. At this crucial point in development, when the sense of time is still loosely formed in the individual, special systems of education could be introduced that would allow the causal field of intelligence to remain intact.

The Necessity of Time Travel

The fictional representation of time travel as an encounter between mechanical mind and human mind is inadequate. The human being is perfectly capable of traveling through time without the use of machines as an adjunct to consciousness. With the advancement of sophisticated computer technology, it will be possible to analyze and study the time/matrix field mechanics necessary to re-interpret and quantify the time strata and prepare humankind for such adventures. Time travel in and of itself is a simple matter. Once consciousness has unified its essential value with the value of time, it has placed itself squarely in the timeless state of that differential context. Time can then simply be trained to stand on end while travel commences.

The human consciousness intercedes with time only when it has the desire to do so. In other words, time cannot be speeded up or slowed down without the permission of the individual or collective fields involved. The human individual who wishes to restore time to his/her own purposes would do so in regard to personal events only insofar as those events might provide representational meaning in his/her field of existence.

The possibility of restoring order to time through such travel appears less imperative once one understands the artificial nature of the time parameters which have been put in place. Once one realizes that there is no clock time and that all matters that are bred into the field of consciousness are universal, then the individual loses the desire to interfere with past events. He/she realizes that all that is relevant to future manifestation is the actual-time, post-now event, the singular moment in which he/she finds the Self.

In order to make the most of this singular moment, one increases the range or scope of perception, thus identifying the significance of that interval. Such preparation is done through a constant state of realization that is manifested through awakened consciousness. The individual plants seeds of the future through cognizing available opportunities, reaching out over the field of time to provide sequestration for future parameters.

Each individual has within his/her own nervous system the full mapping capabilities relevant to his/her personal destiny or time-linear dimensions. The individual must learn to identify this system of intelligence within the span of his/her nervous system and learn the symbolic content that the individual consciousness has used to represent its meaning. These symbols are literal as well as figurative. The individual plants a system of referential time codes through program symbols within the context of the human brain. These program symbols are message units of consciousness which are referential to the primary genetic or DNA structure.

When these time codes are realized through methods of advanced self-contemplation, the individual is freed from the limits of individual field-matrix experience. He/she can restore order through developing systematized matrix systems that are referential to his/her surroundings. This environmental scope becomes available in gradual stages of development, until the individual breaks free from the boundaries of linear experience altogether.

The individual can then travel interdimensionally, becoming free in two respects. First, he/she is capable of developing a much more novel means of intercession with the environment, thus producing an advanced nature of creative perception and awareness. Second, he/she can break free more easily of emotional and physical constraints, thus creating a climate for contentment. The restoration of contentment in the individual is made manifest through a perfect knowledge of the order of time, its intercessionary landscape, and its relationship to Nature. Through this wheel of fortune, the individual can destroy those landscapes which might arise through relationship to a past vehicle and develop a sense of reality regarding his/her present life.

Time as a Messenger

Time must be experienced as a messenger, a harbinger of the created realities which can be assumed through careful intercourse. Time is the only true value with regard to distance and therefore time can travel through any distance simply by the strength of its impressions. The individual who can send thoughts through time can also traverse vast distances of space-time to achieve union with those whose like-mindedness speaks to him/her. Each individual that can relate the field of time to his/her awareness acts as a beacon or message unit for the value of time in his/her sector.

The valuation of thought impulses associated with time is achieved through the development of language. Language in our time frame is made up of stored impressions with response to time, rather than present response intervals. It is difficult for us to contact interdimensional civilizations that do not rely on a memory-based dialogue for communication. We assume, in our time frame, that all civilizations create language through common usage; words are placed in memory banks for use in everyday discourse. However, in advanced civilizations, new language is being transduced and radiated through the practitioner at millions of times a second through the development of thought-process message units that are continually expanding and contracting. It is much like the signaling medium of stars. The on/off central field matrix of communication creates unlimited time values; language impresses itself upon these, creating a type of telepathic union with the listener and/or sender.

This type of communication, which is common in more advanced civilizations, is also possible for us. The reason why it is not more formally practiced is that by its very nature it would destroy the present stance of our civilization. In other words, by switching to a present-time-based method of communication, the human being would immediately be freed of the associative constructs which he/she has mapped out.

The visual/auditory/sensory inputs available to us are vastly underutilized in their present state. With the opening of the time gates, the language experience becomes an artful medium. Much as the artist would not paint the same picture twice, language usage would be changed at every junction point. Each individual or collective member would realize language in much the same way as a painting or a fine piece of music is appreciated now.

With the realization that language is the highest and perhaps most common form of art in the universe, the individual leaps into the realm of language-centered mythology with reference to his/her perception of time-based reality. The person who can play on the field of language can develop the necessary time spread for interstellar travel because the dimensionality of mind/present time reality is unlimited. This allows for an entirely different perception of what life itself is all about.

Time-Spun Language Matrices

When the human individual becomes capable of rethreading language on the spool of awakened consciousness, he/she opens the possibility of unlimited feeling and perception. The array of emotions, which at present is locked into established time constraints, breaks open. At present, the individual is only capable of feeling things for which he/she has a past association. Present systems of psychotherapy spend a great deal of time relearning the behavioral faults of childhood. However, with the breakthrough medium of language restored and a profound understanding of our time-centered reality more firmly established, the therapist becomes a time traveler, instilling the possibility of nonreferential boundaries and breaking the limitations of the mind.

With the consciousness spun out to its true values, the individual can tame the processes of mind and enter into a fuller union with the constructs that create reality. In this way, the individual is united with his/her host intelligence and is free to travel through the manifest destiny of personal understanding.

The relationship of time to language is a complex topic; suffice it to say that language imprints time with its own boundaries. Once language is broken up by time, then it can open to its own interface and create seed memories which are not time-bound. Memory is an important attribute of consciousness. Its function is to restore order to the cellular structure and promote proper regeneration. However, when memory acts like glue in the chambers of the human heart, it inhibits freedom. Memory in its healthy state is a process of inherent rejuvenation; it revives the necessary knowledge of consciousness mechanics. Without memory we would be lost, but without the type of memory we presently have, we would perhaps be better off.

Locator Points in Matter/Time Consciousness

A *locator point* is a piece of the landscape in the intelligence/matter time field. The locator point designates the referential boundary in which the time frame is stored. It acts as a spot in the blueprint for the time codes to intercede themselves and phase in and out. The time frames that interplay on the field of matter must have places through which to thread themselves; the locator points provide the parameters for this interface.

In our environment, the location of a particular object is addressed through a series of space-time coordinates. The locator points pinpoint this inception of time, but rather than creating a singular matrix for this imprint to occur, they expand the matrix through the value of space-time interlock. This quality of expansion, rather than contraction, differentiates the locator points from referential boundaries as we know them. These locator points define the void area by creating a valuable something that manifests as recorded reality. It describes the space between the space that makes up matter.

The locator points are the curved endings of universal origin that have existed since time began. They determine the spiral of intercessional waking consciousness. They are differentiated by clock-time matrices only insofar as they represent Universal time codes of order. Time codes are the holding values for consciousness; the locator points determine the mechanisms through which the holding values will be spun in the awakened field of time/matter/consciousness.

Unlocking the Gates

The holes in the window of the interface are situated through the locator points. They are stretched around the boundaries of time like pearls of emptiness in the necklace of the Infinite. The relationship of time and matter is interposed through gates in the field of time that allow for the transference of information.

Each civilization creates holes in its own shielding apparatus, promoting the possibility of interface with other civilizations. This may be seen on our own planet. When we describe a civilization we often talk about how it interrelates with its neighbors. We talk about its warlike or peaceful quality or the contributions of its goods and services. From a nonreferential point of view, the gates in time are the accumulation of past event-centered experiences of civilizations. They are composed of what we might call the stored goods of past behavior. However, when we understand that there is no past but only a forward present, then we realize these gates are unilateral. They move within their own field of consciousness and therefore exist independently.

Essentially, it is through these portals that we can view ourselves through the annals of time. In a sense, history, as we have come to view it, purports to understand these gates or important intervals in human behavior and seeks to chronologize them. However, these gates can only be understood from a nonreferential or unbounded awareness. When the matrices are revalued in this light, the gates formulate the eyes of the world; blind civilizations are afforded the opportunity to see themselves, thus dressing up their image.

The gates are the windows that the *Host Intelligence* has provided for us to view our essential nature and to open the field for more advanced systems of memory and structure. The realization of such gates in the memories of the Earth continuum must come about for us to re-realize our identity as a species.

The Marketplace of the Infinite

The establishment of pure consciousness as a breakfront reality in the human nervous system promotes the development of non-boundaried existence in which the individual can lose him/herself in the marketplace of space-time cognition. The wares are the stored references that have been sampled over the individual's entire referential landscape within the Host identity. In order for the individual to bring him/herself back from a state of forgotten memory, he/she must recognize the inherent valuation of memory with respect to his/her travels through space-time.

The memory that must be reestablished is the nascent structure of interaction between what we would term the personality and the soul. Each individual must rediscover his/her eternal matrix by returning to the central marketplace—a journey into the original light. Each individual must stand alone in this journey, but it does not need to be a struggle if the understanding of pure consciousness is made available. Once the individual recognizes the initial steps necessary to identify the birth matrix, he/she can continue to recognize them as they re-emerge at any given point.

One may think of this process as a pairing-off of locator points or event chambers in the unconscious memory. The individual reconfigures his/her past events so as to break free from unending repetition. The individual gives up reference to his/her past, leaping over the back of change. Original tonalities which were created by the Host Intelligence can now act as reference points for transdimensional life. A non-event-centered personal reality becomes possible. From this position, the individual can be birthed into new beginnings from every point without disorientation.

The personality is simply a representative of infinite consciousness. It acts as ambassador for the Host matrix and determines which advantageous steps in the nascent blueprint will be made available. The soul value holds steady, waits for the proper interface, and infuses itself directly onto the projected matrical field when properly invited. The soul holds back, not becoming overly deterministic, but waits for the proper timing to transduce the imprint of full awareness directly to the individual.

Loosening the Hold on the Personal

The personality can recognize advantageous material which will be profitable for its growth. It does this through the laws of attraction which precipitate organizational matter directly from the Host Intelligence. The personality curves over itself in reference to its own idiosyncratic nature. It develops characteristics which form a prism through which subjective consciousness may be displayed. The characteristics of the personality are developed through the soul's determination of timing.

Individuals whose personal style appears more fixed differ from those who appear more flexible through the process by which they interface with time. The looser individual can release pressure through proper give and take with the environment. Individuals who are more tightly wound have difficulty permitting the flow of pure consciousness. The tighter individual maintains stability through fixity of mind rather than through heartfelt self-examination. The personality is nourished by pleating the mind with continual reference points from other strata of existence. The mind looks for relief through the pursuit of emptiness. Gradually, we realize the value of a completely nonreferential identity.

The jump from a chaotic mind to one that seeks emptiness is a positive step; it leads to the fullness of a mind that can function through a personal context that needs no fixed boundaries for stability.

The Personality and the Persona

The individual who jumps into the existence of pure consciousness as a precognitive value makes room for the persona to begin to drop; the primary time values are stirred up. This individual, now somewhat free of constraint, can choose reference points from a more infinite basis.

This habit becomes habit-forming. The mind develops independent reasoning. It does this through pure cognition which by its very nature changes its point of reference continually. The locator points are shuffled through the play of the infinite; dimensionality occurs in parallel terms.

The individual can then create simultaneous seeds of endeavor in many dimensions; in each, isotopic blueprints of developmental order can be constructed. The individual fans out his/her influence through this *transdimensional* experience and strips away the relative persona that has bound the personality. Such a transformation restores personality to its full meaning—a coat of arms for pure intelligence to play out its drama.

The individual, though maintaining characteristics of personal identity, is freed from being consumed by them. He/she can live a value-free existence based on the time-dominant wave mechanics of personal distinction. Small, unique brushstrokes are characteristic but yet non-limiting. It is as if a painter could paint reality through the congruency of a recognizable style; the painting would be unique in every other aspect. Sameness remains but differentiation is highly visible.

Full differentiation, combined with a unified field of expression, opens the individual to an increasingly original perspective on reality. The person then has the opportunity to become truly healthy with respect to his/her development in time.

Release of the Personal "I"

Each individual must shed the personal need to express the relationship of his/her own value to the Infinite. This is a difficult and risky undertaking for most, because they sense this as a shedding of their defenses regarding the building of identity. However, if the identity can be seen as a mask, which when pulled aside gives forth to the full rainbow of cognition, then this undertaking need not be viewed as a diminution of human activity.

We involve ourselves in a push-pull process in our group dynamics which implies that one person must be dominant and another be submissive. This is the collective bargaining we are accustomed to in our search for the "we" value. However, when we look at the "I" as a matrix (by definition filled with a myriad of "we" values with respect to time, matter, and consciousness), we do not have to feed additional subsets of possibility to our awareness. With infinite time as our vantage point, when we merge with the collective "we," the personal value leaves behind a trace of something new—a pure intercessional time response that differentiates one unit from another. Though subtle, it is the relational pool from which individual intelligent species are drawn. In other words, the Host Intelligence asks us to draw upon our full experience as a differentiated identity until such time as that differentiation impairs our progress. At that point we are asked to come back together again, leaving behind that which we have known as ourselves.

In this process, the "I" value becomes completely indistinct from the pure time matrix. It becomes a corporate entity of time, circling through the expanse of consciousness and opening its eyes to the field of infinite value. In this weightless, motionless, effortless state of awareness, the "I" value enjoys itself immensely. It is free to travel through every distinct corridor, knowing full well that it lacks this distinction now, but not missing it. It is completely full with the "we" and does not need to be braced up by its own need to have more.

In this state, which is transitional, the "I" value is weightless, but it maintains a separate status. It is joined at the hip with the referential but it is free to move with the infinite.

Transiting to the Full "We" Value

After the "I" has been allowed to play, to establish its sense of nonbelonging, it can trip back over itself again. It goes through a period of God-differentiation in which it experiences itself as all of the values of infinite realization. In this state it is part of the Host Intelligence, breathing its breath, but still distinct in its intelligence matrix.

In this manner, the "I" lives alone until it can build its own house in the natural world. When the "I" value grows tired of play and seeks the reaches of infinite living, it actually strikes out on its own. In this pose, it is at once completely free in relation to itself, completely cast off from its former origins, but it is more unique, not less, in its final outcome.

When the "I" becomes the true "I" it is the eye of God itself. It has no reason to fear. It is completely taken up by God, yet it has transferred the Host Identity into a full matrilinear value construct. It perceives itself as separate, yet it is actually a part of a counterpoint to the living presence of God. In this state, it is value-transparent with respect to its ability to perceive reality. It is still capable of individual creation, but its individuation is always and continuously a manifestation of the One Source, of the one infinite nonbinding reality from which it has stemmed.

In short, it lives and breathes time and becomes the relational value for time to imprint itself on the cosmos. The intelligent individual becomes the facet through which time can express itself fully, without impairment and without having to refer back to Nature. In this state the "I" becomes the casting booth for Nature, directly opening the range of environmental liveliness into the individuated human experience. The individual becomes the entry point for new consciousness values, a bed in which new seeds of the Universe may be planted. This is a most exciting, valuable, and interdependent state.

Beyond Individuation

In the quest for individual blanketing of perception in the Host Intelligence, the individual creates a span of unified constructs through which to express his/her individual reality. This circumstance is necessary because it posits individuation; however, when the field matrix is broken up through the new valuation of time as it is impressed in the field of consciousness, the medium state of referential being returns to its nascent value.

In this sense, the individual becomes a breaking-off point for the pure consciousness relationship. He/she enters a dormant state, storing up time values for possible curvatures into the space-time vault. The unification of reality, which occurs when time meets up with space, is the vantage point that all individual intelligence matrices must use in their quest for infinite realization. However, in the case of our species, it is now time for us to catch up with ourselves.

All of the vantage points that we have interposed to create prerecorded history in our context have been speeded up so that we may catch up with ourselves in the developmental field. We are "chasing our tails," waiting for the right time codes to present themselves, so that we can skip ahead onto the field of direct cognition. By impressing ourselves with the possibility of infinite absorption into the field of unity, we are scraping the bones of our previous existences. We are manifesting a terminal reality, a one-pointed, intergalactic field of possibility in which we are the galaxies, and the cosmos is our region of operation.

The "I" value, seeking to destroy order through magnifying the range of its own perception, will be swept away. In its place, the human species becomes a "we" matrix, with the "eye" being the central focus of realization. It breathes a sigh of relief as it makes its way into wholeness. There will not be a loss, but only a stretch into the principal identity. However, as it occurs, the psychological parameters may be confused a bit, or even frightened. It is difficult to pose a "we" variable to an "I" parameter.

It will take time for the value system to be stretched enough for the hammock of life to be draped fully over the construct of space-time variables. The lifetimes of relationship we have accumulated are being stripped down to accommodate our new viability. We are being invited into the substrata of universal reality, but not before we have been fully tested as to our breakability in the field of the infinite. We are time-tested but not quite ready to enter the marketplace!

Breaking Free of Time

The curve of space-time affords the possibility of meeting the junction points between consciousness and matter. As the curve realigns itself during the next twenty-year period, we will be crisscrossing over the time matrix and entering a more balanced and unified sensibility. The relationship between consciousness and matter is stretched apart and identified as to its field reference constituents. We will enter into this relationship through a period of grace during which we will experience the interstices between ourselves and the environment.

Once time is relocated in our awareness and allowed to break free from past constraints, the actual tenor of matter will be afforded the opportunity for change. The pitch of matter is determined through its chord of entry into the portals of freely expanded time. Since time is infinitely stretchable, it is also infinitely breathable; as time breathes its own nature directly into the field of consciousness, new forms of matter are the logical result.

Matter cannot continuously expand into consciousness without periods of uniformity. It does not freely regenerate or differentiate; it returns to its undifferentiated and unmanifest value for regrouping in the field of space-time and distance. Once matter has recalibrated its position with respect to consciousness, it can leap forward into individuation without stopping to consider the time values. The time values become internally mirrored in the essential nature of matter.

As new forms of matter are created, human beings will have the light-bearing opportunity to interact with their environment in a new and more retentive way. They will be capable of ingesting the memory of structure directly from their environment and recasting the boundaries of space-time.

Creative Time

Once time has reestablished itself, through the break in the fields of reference between matter and consciousness, it will be able to run continuously. This run-series of time, distributing itself through the interplay of space-time, will be able to be calibrated through our effortless movement towards spontaneous creation.

With time open to the field of expression, the balance of power between the dual values of light and dark will be shifted as well. In a non-dual value, consciousness seeks its outgrowth through relative sameness, conspicuous maturity and internal graphics. It seeks the interrelated fields of its expression by reaching out to the internal movement of its own breath and establishing roots of enterprise that go beyond its previous evaluation. Constantly making its constructs more spherical, time can rise to distinction from the bed of matter. Seeking union through enterprise rather than having to duel with the boundaries, time is free to compose itself, to establish its relative causality without battle or conflict of any kind.

There cannot be any true world peace until the need for duality ceases. It is felt that this cannot happen until the nature of matter as a clock-time experience is altered. Until matter can be completely free to express itself through infinite correlation, time will constantly wind its ribbons around it and cause matter to break out into the field of opposites.

Therefore, the best healing for time itself is our own knowledge of the infinite roots of matter and the free-wielding value of consciousness in our own internal and ever-changing reality. When we become less dualistic and more interwoven in our expression, we will no longer need to interpattern our life experiences through conflictual representation in our own universal constructs. We will destroy the boundaries and thereby prevent the risk of destroying ourselves. This is the benefit of a retrofitted time value, interdependent, valued in consciousness and garnered in love.

When time is freely spoken, it always speaks the truth. That truth, vested in kindness without effort, is the manufacturing value of which our soul's awareness is capable. The intonality of time, stretched breathlessly through the icescapes of our formulation, free, interlocking and manifested without end: this is the destiny of our planetary configuration and it is the grace of time.

CHAPTER TWO

THE PSYCHOLOGY OF GLOBAL RECEPTION

CHAPTER TWO

THE PSYCHOLOGY OF GLOBAL RECEPTION

Uniform Dimensionality

The progress of the human species depends upon our ability to scan the field of perception, developing constructs that are adaptable to new modes of living. Each of us maintains a core matrix of sound and energy which provides for creative advancement. The ability to shape the field of reality enhances our global receptors, bringing us closer to full realization.

The global mind develops itself through the prismatic interplay of perception. We approach this perception through psychological or psychobiotic terms. We imprint our body/psychology with a mediating context for our perception, developing blueprints that determine our principal desires. A new scheme of relationship would allow us to draw on past experiences, scanning them like a diary but without drawing conclusions. By correlating every scan through a uniform set of response media, we embrace the full range of dimensional interchange. This gives us the capability of bringing absolute life to the field of relative existence.

Uniformity in this context does not mean sameness with respect to form; it means sameness with respect to the depth of origination, the controlling mechanism of life force or activity. In the development of higher states of consciousness, the experience is that the source of divination is connected to a greater chain of being than we were used to. As we gravitate towards a deeper realization of sameness, we appreciate the reflection of infinite diversity.

This golden paradox allows us to live every event in the moment of its inception, while stretching the range of motion with respect to its interplay in all fields of life. When the human being is inherently balanced, he/she is able to *free-gap* perception, lifting onto this infinite field even as static perception remains. In this state, he/she is onto the mode of existence we could define as uniform.

Interplanar Psychology

Our present understanding of the mind is based upon our ability to glimpse relative pools of perception, thereby attempting to describe the source points for consciousness/apprehension. The psychological constructs we have developed seek to explain the rationale behind human awareness and behavior. We are now at a point at which the matrices of time afford us the possibility of developing an intermodal psychology that is both disjunctive and conjunctive. We experience the mind as a habitat for integration. The mind seeks to unify even as it seeks to split things apart. Psychology can address these seemingly opposite tendencies through creating therapies which examine the underpinnings of the psyche while emphasizing transcendental unification.

The new psychology attempts to change the fabric developed through present lampposts of reality, while identifying new locators in the field of infinite perception. This type of psychology would strip us down at the same time as it would build us up. It would free us from all qualifying media. Conventional attempts to explain the origins of a given psychological framework would no longer be viable.

The quest for privacy in the human mind is determined by a childlike desire to control the playground of the impossible. This new framework, which cannot be controlled, will be soft at first, as the individual or collective is not capable of storing all of the information necessary to develop the full range of vision. In practice, therefore, we need to live in a field of ambiguity, realizing that we can no longer return to a definition of Self that is based on cumulative experience.

At the same time, our stretch is not yet full enough to embrace the infinite. Although our place in the field is not fully defined, we have the advantage in that once the mind has been identified as the modal unit for consciousness, it can be hitched to the wagon of realization. From this point, it can seek the ultimate goal—the manifestation of a simple interchange with the breakpoints of unqualified perception.

This transitional position helps to redefine psychology as a realization/response matrix, rather than an analytical process. Instead of reducing ourselves to the small events of life that have occurred through the gateway of our own personal panorama, we can take the blinders off. We then understand that the mechanics of the nervous system that catalogue memories are not memory itself. Pure memory, which is steeped in cognition, is held to the glass of perception for distinct intervals.

Pure memory is distillate consciousness, pure and refined and not mixed with the colorful emotional jargon we have heaped upon it. The new psychology will develop a place for emotion by identifying it as the subtle medium through which realization expands into the field of life. When each chapter of our lives can be realized through this valuation of emotion, expression becomes uniform because it is bathed in the absolute. It cannot conjure up any hidden ghosts because it is knowledgeable of itself, responding directly to the field of activity in which it finds itself.

Psychology that is wrapped around collective constructs limits expansion because it must represent cultural values through every field of activity. In this sense, "psychology makes the man." It is probably scary to think that psychology could unglue the hinges of advanced perception, allowing us to skate freely over the surface of accomplishment unhindered by any fixed moral standard. This is because we assume that cultural individuation holds in check our abnormal tendencies.

We greatly fear these tendencies, which are self-violent or deterministic in character. This is with good reason, since we have not yet come close as an entire race to realizing the essentially powerful nature of creation. We have set artificial goals and resistances, embedding them in the webwork of our psychological functioning, so that we can count on being accessible only to our more noble instincts.

However, the rules we have established are intermediary. Though they may initially have been necessary to us as thinking/feeling beings, we now find them in our way. We must learn to function from the level of cognitive realization rather than from imposed analysis. The goings-on of the psyche, the "rap" we generate through the thinking mind, clouds the vision of *early-time breakthrough*, leaving us gasping for air.

To correct this condition, we are being asked to reidentify our interplanar constructs of existence. We are being asked to develop "quadrary biceps," muscles of motion that will stretch our perception into the quadrants of unidimensional reality. The transition from aspirant-seeker to psycho-spiritual adept involves no particular dogmatic formula. What it does involve is a willingness to leave behind all of our given truths and to seek a refinement of sensibility so that cognition can sing its own perception.

When we can instinctively guide ourselves by linking our unimodal personal selves and our multimodal interdimensional selves, we will be on the way to perfection. The link is between relative causality, which expresses itself through identification of the chains that bind us, and a new system of carving up the boundaries of time so that our interplay can be cosmic in the best sense.

When we curve back over the foundations of character, the psychology has room to grow. It can expand into an identity that is humorous, functional and musical. It represents the character of identity as its tonal dominant. Psychology becomes the study of the naked value of perception rather than of the mechanisms through which perception is gained.

The Value of Feeling

The roots of emotion lie in the heart. The heart is the principal repository for the distillation of pure feeling. Each strain of pure feeling in the human being has a positive or higher energetic outcome that is produced through the refinement of that aspect of the central nervous system.

When the individual develops a field of response to a given situation, it is usually approached on two counts—intellectual and emotional. The intellectual response, which has the predisposition towards analysis, can bend the truth, because the capacity to reason the outcome of a situation often leads to unreasonable conclusions. On the other hand, the capacity to feel one's way through a situation can lead to outcomes that are predisposed to disunion, impulsive behavior and strategies that are unconscious in their motivation.

The skill of operating within a continuous field of refined sensibility, which is both reasoned and emotional in its approach, needs a different point of origin. When one looks at a given point, one is seeking to determine the character of relationship between the situation and oneself. This character of relationship identifies the point in the field in which one resides. One is navigating through the terrain of thought/matter possibility.

In this situation, if a person can enter into a meeting with the territory of consciousness which is to be crossed, a relationship of substance can occur. In other words, the individual can merge with the vibrational resonance of the characteristic environment and become attuned to or aligned with it. In this state of awareness, the individual leaves behind his/her opinions, or even his/her feelings with response to the material at hand. He/she enters into a constitution of floating or meeting up with this decision/question in such a manner that consciousness meets directly with consciousness. In this sense, the individual becomes part of the entire matrix of possibility and for one brief moment may in essence be floating in time.

Through the experience of intense feeling rather than emotion, the individual can profoundly sense the essence or character of a situation without being gripped by it. This manner of functioning begins to translate itself to all fields of life once it has been adequately established in one area. The quality of obtaining an experience of life that is feeling/felt rather than intellect/poised is essential for full development of either faculty. The intellect caught in a method of interaction that develops decisions and analysis based on past data is always limited to the past data to which it has access. The person who exists in a state of constant emotional upheaval or frenzy cannot locate the seat of compassion in his/her own inner life. For this reason, there is a great need at the present time for individuals to develop unity at the feeling level.

The Essence of Pure Emotion

When the seed of consciousness is planted for a given idea or aspiration the emotional content is stirred. No point in creation is established without this measure of feeling. It is integral to the nature of life itself. It is not only our response to the rose that is emotional, the rose is itself a product of the feeling value of life. It is representative of the kingdom of roses and has within it the medium of speech for that given substratum of species. Every important valuation in human activity has within it a seed emotion, the subtle range of feeling that expresses the character of that action, event, or material creation. As our perception becomes more refined, we become capable of merging with that field of perception and can literally feel the rose as we can feel the presence of someone we love.

This feeling value becomes more and more pronounced, even though the intensity of emotion in response to that event may level out. In this way we become constantly sensitized to the environment without being overshadowed by it. This pronounced intensity, which is at once very pleasant and at the same time neutral in quality, develops until the full range of bliss or overt union with pure consciousness is fully established. Emotion becomes a playmate of diversity by expressing itself in the field of activity as a sense of wonder, awe or reverence for what is taking place without being submerged by it.

This process of resisting the overwhelming is at first strange to us as we develop psychologically, because it seems that to be lost in the feeling value of our sensibilities is a measure of our humanness. As time goes along and we register our qualifying sensibilities on a larger scale, we are free to experience a wider and more pronounced level of feeling. We are not limited to the stronger tastes which block the subtler tastes from being experienced. Our entire experience becomes more diversified.

Seeds of Expression

As our ability to perceive distances in the relationship of time and space begins to grow, we are able to understand the efficacy of reaching into the pocket of the infinite in order to expand the feeling level of our own consciousness. We continuously seek out new experiences to press us out and establish new boundaries for expression.

This dynamic interplay generates a response to activity that is uniform in causality because at every junction point it is revealed that that we are matter and matter is us. We are unable to differentiate between what is and is not ourselves. Our perception becomes uniform because it does not make a distinction between ourselves and the environment. At the same time, it is fully appreciative of every aspect of existence from the level of perfection. The mirror reflects the nature of reality back onto itself in a constantly expanding reality formation.

The emotions are the control buttons of this infinite expression. The emotions create memories of seed feeling that cause us to remember what we are when the boundaries between ourselves and the surroundings are not so distinct. Of course, this is a process; in this process the entire curve of function between ourselves and that which we know as "other" becomes less distinct.

This can be felt most strongly in interpersonal relationships. In a romantic relationship, our usual response mechanisms cause us to experience ourselves as either separate from the loved one, or so completely subsumed by him/her that no distance can be achieved. Once this distance is lost, we interfere with time perception and are left holding the bulk of time in the context of the relationship. No new avenues of interrelational activity can be established. We are glued in to the superconstructs of past/present/future without recourse to originality.

However, in the new perception, the individual views the other as simply an open medium through which thought/emotion response parameters can travel. We view the other as ourselves, yet in transition, we continue to appreciate the diversity that makes this person distinct from ourselves. In the dance between other and same is a movement that reaches a point where reality becomes emotionally uniform. We experience the range of feeling that creates bliss; the feeling of perfected expansion is realized by the union of two people who remain free from constraint.

This experience of encapsulated unity, which is the first experience that we as human beings are apt to have of an awakened field of consciousness, is marvelous. We can drive the bus home towards our own happiness, and feel as if we have some measure of control. This is because we have not completely realized the full interlinking of consciousness on a wider field and in a sense still have some measure of causal destiny.

As emotion becomes distilled, it becomes very subtle in its expression; so delicate that the slightest tilt can throw off the direction of its approach. Time must stand still in order for emotion to be experienced in its delicacy. Time must literally trip over itself and remain rooted in one position, so that this quality of feeling can be expressed.

Time is friendly to the establishment of subtle feeling in manifest creation. In its partnership with Nature, time has the responsibility for supporting the efficacy of feeling; this establishes the notes of activity for matter to form. The seeds of matter are created through this subtlety of feeling and this is why civilizations that have developed subtle intense feeling also are capable of fulfilling the manifest creation of matter.

The question of how feeling may be both delicate and intense is posed since these two responses seem like opposites. Once delicacy is established it spreads itself over the field of reality so that it permeates every aspect of identity. In this sense, the fragrance of love is an actual reality. The expansion of bliss through the physiology of the individual or collective becomes a band of pure gold over the response mechanisms of the individual. The person is literally coated with love and cannot escape from it. The field is firmly established and the locator points in creation can then come forth.

The feeling expression for the flower is the actual manifest map for its creation in the field of matter. Every stage that has been described in the development of matter has its corresponding field mechanics with response to the perception and range of feeling. Therefore, to experience or objectify matter as something which is cold to the touch is inaccurate and perhaps destructive to the development of human sensibility.

Cross-Cultural Patterns of Emotional Union

People usually unite with someone with whom they can feel some measure of sureness and thereby a greater measure of security. This allows us to cross into areas of development in which subtler levels of feeling are available.

However, if we hold on to sameness as the only criterion for union we limit contact with other races, cultures, and ultimately other species of intelligent life. Therefore, we are being asked at this time to leave behind our concepts of union and develop breakthrough unions with people whom we would not usually meet. This involves leaping over the bounds of convention and establishing relationship with people of a different color, a different perceptive slice, or who are simply different in some profound but challenging manner.

This accomplished, there is a crystallization of consciousness that promotes the individuals to heal old scars of dissolution that they have been carrying through their existence. Each individual reframes the genetic conduits that constitute his/her landscape and enters into a creative dissolution of the bonds that have limited personal and collective freedom.

To bring about the full flowering of our racial expression, the taboos that have been set in place regarding crossover in marriage or relationship between members of divergent cultures or racial backgrounds must dissolve. Racial integrity need not be bound to maintaining sameness with regard to racial interaction. The purity is in the Self. Collective realization of core identity for any group or subgroup of people generates understanding of its origin of purpose in the manifest creation. This level of responsibility is brought about through devoted contemplation of the source of identity, with proper training, dedication, and effort.

All of the ranges of feeling that are possible for humankind are to be found in the range of racial expression that has manifested on our planet. As we experience people from other cultures, we are treated to ranges of feeling that may or may not be present in our own constructs of reality. By limiting ourselves to those who are like us, we cannot truly find out what we are. This develops a strain in the field of open expression that does not allow the heart to fully develop.

The Mechanics of the Heart

The heart is not merely a symbolic representation of the province of feeling in the nation of the individual. The spiritual heart center is the repository for the biochemical mechanics of resonance that allows emotion to be encapsulated in the human identity. The spiritual heart unifies with the physical heart to form a conjointly matrixed network of light/energy/sound that manages the intercausal links between mind and body. The physical/spiritual heart matrix calls up thought patterns from the mental field, identifies them in the feeling tonality of the heart and then recorrelates them so that they can be fed to the nervous system for repatterning. The nervous system sorts these signals, establishing modes of conduct that redefine our sense of reality. Our difficulty is that most of the time we are used to interrupting the flow of these signals before they have been fully integrated into the body/mind.

We allow our thoughts, which are actually just a link in the system, to become the defining centerpiece for our understanding of the world. However, when the signal system is allowed to flow smoothly between the physical/spiritual heart and the mind, the nervous system can fluidly conduct signals to the physiology that allow for spontaneous right action. As this process matures, the mind becomes silent, bypassing what may once have seemed to be relevant processes and using energy to restore order and mobility to the heart. The heart is no longer caged, but can inspire creative behavior without being constricted by past limitation or fear.

The physical heart establishes itself in this domain through uniting biodynamically with the human nervous system. These chemical links, which have yet to be fully studied by science, represent the natural mechanics of functioning which govern intelligence.

As we think, so we feel. As we feel, so the heart produces chemicals of feeling that are very specific in their content. The heart literally produces hundreds of thousands of chemical corridors through which the receptor mechanisms of feeling can travel. As the heart develops, the range or networking system of feeling/perception becomes more detailed and also more efficient. New chemical circuits are laid in which allow us to interlay feelings one on top of another.

The heart is a layer cake of time values. The heart manifests feeling conduits through which time/emotion can travel. Time establishes itself through the biodynamic clock of the heart and, in so doing, allows the heart to create open fields of activity through which the response variables of the environment can make contact.

Each environmental approach involves a different stream of union within the heart. As the heart develops more fully, every variable is experienced as more united and at the same time more diversified. The environmental trigger opens up a new pathway through which consciousness can travel to reemerge at the source.

It is as if the heart is the emotional traffic cop of the identity. It allows the individual or collective to beat on the drum of the infinite and establish the rhythms that make pure consciousness become individualized on the level of the body/mind. The heart provides the orchestra of awareness for consciousness to occur.

When the heart has been damaged, either physically or meta-emotionally, "ice" forms on the crystalline structure, and the contact points are not as clear. It is as if the individual becomes confused as to the nature of feeling and can no longer distinguish what he/she is made of in relationship to his/her contact with the infinite. When this occurs, the individual is liable to close down and in essence cut off the field of biodynamic energy that makes life possible.

Heart Flow

The individual stores emotion in the spiritual chambers of the heart. These chambers can be viewed as inflow and outflow units which govern receiving and giving respectively. The spiritual heart chambers, when in balance, allow a flow of divine energy from the higher spiritual centers to enter the heart, creating a feeling of spatiality. Through these subtle interfaces, the heart scans the field of love and records impressions of psycho-emotional intimacy. Through meditative techniques, it is possible to resequence the flow of energy in the heart and thus free up the energetic pathways. Once the heart is established in expanded consciousness, it can travel at will through all of the dimensions of activity and can develop its own soundbed of intelligent feeling which is transmitted to the nervous system through biochemical interlay.

Techniques used for restoring heart flow help the individual to be sensitive to the province of feeling within the heart that is perceived to be connected to his/her presenting issue. The Higher Self is engaged to witness the areas that need attention. Primary emotions, such as anger, sadness, or fear can be felt/sensed in the spiritual/physical heart matrix; as these sensitive areas are mapped they can then be recalibrated through the introduction of inner insight and Divine energy. The individual finds that he/she is gradually able to create different modes of response to life situations, and is no longer involuntarily governed by past trauma.

The individual begins to identify the subtle centers through which energy circulates in the spiritual/physical heart matrix. He/she finds that by centering the attention at the base of the heart, a feeling of strength and resilience is achieved. This understanding, when combined with a meditation practice that fosters silence, can guide the individual to an experience in which he/she becomes rooted continually in the heart's reality, which retains simultaneously a dispassionate witnessing character and profound feeling. The individual finds that with the signal given to return to the base of the heart, he/she can trigger the biochemcial matrices that direct the entire body/mind to identify, restore, and in some cases reinvent new energetic pathways. The individual learns to live life centered at the base of the heart, directing the mind/feeling synaptic interlink through the depth of spiritual transcendence. When pain of either an emotional or physical nature is introduced, the body/mind learns to direct its attention to the base of the heart and to center the attention in this region. The heart/mind that is created can simultaneously think and feel without being overshadowed by either function. A general atmosphere of equanimity is attained.

The heart that is functioning smoothly is capable of developing a range of feeling that is so easily flow-soluble in its continuum that each tiny perception of feeling is tracked onto its own biosynaptic pathway. The rivers of life running through the heart are gated, directed, and developed so that the individual can begin to experience very subtle ranges of perception. The heart becomes soft, expansive and at the same time very definite in its readability with relationship to biochemical ingenuity. The heart can retrace its steps in the infinite, storing data for future use and developing very complex networks of interaction between itself and other forms of intelligence both inside and outside the individual.

The environmental networking that the heart develops frees it to travel to other territories of perception, causing us to feel close to others. The expanded heart actually travels to another human being through these subtle interfaces, scanning the field of love and recording impressions of psycho-emotional intimacy.

Evolution implies a constant, dynamic laying down of matrices of formation as directed by the heart and as assisted by the timing mechanisms afforded through proper circulation in the valves. When we lose touch with reality during trauma or injury the heart literally becomes confused and cannot identify its own range of perception. People who have weak hearts for any reason on a physical level may therefore be plagued with a sense of doubt, insecurity, or fear. The biochemical strength of their own heart matrix cannot function and therefore they are left questioning their existence rather than enjoying it fully.

The Divine Heart

The Supreme Intelligence or God Consciousness calls to us through the interlay of the heart. As the heart expands, we are able to bridge the gaps between ourselves and the full cosmic light. As this occurs, we literally merge with the heart of God.

Experiencing oneself as a silent, motionless unit of God, the individual can maintain the dynamic capability of full motion and activity. One finds that the restlessness which is so much a part of human nature is instantly and permanently reduced. No longer needing to hold to a fixed value of time as the basis of activity, the human being whose heart is fully evolved is free to develop a meeting of time through the true psychology of the Infinite. In this nonmoving, almost breathless state, the individual develops a lock on time and can live permanently in this clockless, infinite value. One views everything as continuous, all present, ever-altering, yet completely changeless and perfect. This experience of an expanded yet static universe defies description, yet it is the heart of the heart.

The heart, in a sense, is symbolic of our universe. The rhythmic breaking of the waves of the heart, its constant, steady beat towards the infinite, mirrors the entire cosmos as we know it. The expanding, layered universes of light are traced within its own physical walls. When we view the heart as the mirror of infinite construction, then we understand that to develop its possibilities allows us to fully embrace manifest Creation. For intelligent life to develop we depend upon a biodynamic interplay of love values; the heart is the feed mechanism through which such an understanding can be gained.

The loving individual who has lifted him/herself into the domain of cosmic consciousness can view the universe from a place of finality and expansion at the same time. Each breath is lined up directly with this motion, and it is perfectly synchronous with it. This lubricated movement of consciousness gradually produces more and more refined states of bliss. The spiritual traveler need never get out of his/her seat, yet can view the domains of life from every angle. With this full range available, all possibilities are apparent. The heart leads the way to a synchronous, uniform consciousness, while constantly manifesting subtle forms of interlocution with the body/mind.

This type of heart, which is healthy, evolving and mischievous in its conduct, is the type of spirited vessel we are looking for. It is at once perfectly self-contained and yet constantly seeking partners in the vast continuum. It is never satisfied, yet can reach an optimum level of contentment in which it no longer needs to identify new frontiers of feeling through interfacing with its environment. Once the heart has achieved a state of homeostasis on the level of feeling, it seeks to reach out to the infinite without seeking restoration.

Psychological Approaches to Meeting Time

Since time is not a constant, but by its very nature is continually changing character and shape, the individual who hopes to capture time must change shape with it. This leads to an understanding of what we might call *shape shifting.* This concept implies that the individual will lose form as the rate of evolutionary spiral is speeded up. This is precisely true. Shape depends on a unity of form, and form cannot be constructed out of disparate matter which is constantly moving on edge. The individual who seeks direction for his/her development is looking towards breaking the doctrine of form within the field of infinite structure.

The human individual is composed of layers of interwoven light that constitute his/her form in time and space. These layers can be rearranged by interlocking with the fields of time themselves. It is like running a race; just as one nears the end, there is a feeling of being at one with the road. There is a sense of effortless timing, of coasting along on the sea of the infinite in a frictionless and expansive way.

Individuals are now evolving towards a non-static body mechanics. We are beginning to experience ourselves as fluid rather than solid. We can repair ourselves through the expression of our own consciousness and can literally lay down new tracks for ourselves in awareness. Our humanity is not based on what we look like, but rather what we are.

When we begin to change shape in relationship to how we think/feel/behave, we take on more of a free-form matrix of activity with respect to our environment. This goes beyond the notion of endless time or individualistic non-aging, but in fact stretches to the notion of a body mechanics that is bioenergetic rather than biostatic. The individual who can begin to supercharge his/her body with the living consciousness of memory can redraw his/her figure and develop a range of freedom which was not previously possible. Although this may seem like mere fantasy to us, there have already been humans on our sphere who have lifted free from the lock-in of the body/time hollow. They have discovered the fullness of expressed consciousness not bound by the skin.

Skinless Time

Time encapsulates itself through memory and form. These are the mechanisms through which it holds itself back from recurring again and again. The human form is an encapsulation of the psychology of time. Each human individual is a storage unit for the soul memories of that entity's travels through the Infinite. The soul itself is a complex message unit that records the fluid motion of time such that it can be reprinted in many forms, over and over. This self-replication value of time allows for the development of species and for characteristics that identify those species.

Souls, however, are not species-specific. Is there something specific to a human soul that makes it human? Each human being is composed of a matter/time ratio that allows him/her to be whole in the corridors of time. The human soul differs from the nonhuman only in its ability to fully cognize its own nature. The possible expanded values of cognition of which a human soul is capable distinguish it from other forms of life on this planet and perhaps also from other species of life in this dimension or beyond. The soul drinks the tea of forgetfulness with respect to its balance in time so that the human being will remember itself only as the form in which it presently resides. The soul is the core energy unit that constructs matter to form a living creation.

The psychology of time allows us to look beyond the living matrix from which we now operate and build others in which we can escape once it is right to do so. Our soul's development depends on our willingness to call forth the full range of motion from our immediate selves and to cast this off into the infinite. Our humanness is dependent on our willingness to change rather than remain the same.

It is in the interface between static and changing mind that humanness can be identified. As we change, we increase our range of feeling, and as our range of feeling expands, we can interact more directly with different strata of existence. Once our dimensionality is uniform, we can be identified as human through our specific feeling-range in regard to nonhuman species.

All life is intelligent. It is the anatomy of that intelligence with respect to its manifestation in matter that seems to identify its nature. However, our spiritual anatomy is based more on our willingness to be infinite than our tendency to remain enclosed. As our consciousness releases the time capsules stored in our soul's nature, we will be free to travel throughout the corridors of time. We will identify ourselves as human by our ability to risk the full embrace of love. This love, which is divine, ever-radiant, and non-causal with respect to matter, embraces all of the dimensional mountains and climbs them step by step. Our evolution as a human species depends on our willingness to climb these mountains and to seek a heartfelt response to them.

Psychological breakthroughs cannot be stopped once they have started. With all of time as the launch platform, psychology can offer itself as a vehicle for consciousness to express itself more competently. The outcome will be that true psychological understanding will bring us to a place from which we can fly on wingless expanses of light. That is our destiny.

Psychological Infrastructures

Our concept of mind refers to the contents of the box rather than what lies beyond it. We identify ourselves with the crossbeams of intelligence, assuming that the parts which intelligence can grab onto present the entire spectrum of identity.

Psychology is the tool that consciousness has developed for us to experience perception. The mind is the storehouse of intelligence, developing figures for the intelligence to follow, much like the classic lines of a skater. The perception skates over these figures, developing forms to which reason can cling. How would a new mind function? Would thinking occur in the same style as we are accustomed? What are the lines of force between the mind and pure intelligence in relation to opening ourselves to the tree of knowledge?

Psychology must lean into matter. In other words, when the psychology is trim in relationship to consciousness, it develops open structures which are smooth, prismatic, and developmental. Intelligence then leans into the consciousness, developing still forms as well as changing forms.

Consciousness needs both changing and nonmoving lines of feeling in order to develop the perceptual bands that create mind. As our mind develops, it creates principally out of nonchanging variables. Rather than a stagnant field, filled with old forms upon which we can draw, the mind creates pools of memory out of absolute consciousness. It analyzes the inner dynamics of its own matter and reveals all of the drawing lines through which consciousness can expand.

Upon this analysis, the mind reaps matter directly from consciousness. The slope of matter is drawn and the mind recalibrates itself so that it can draw matter directly from the curve of consciousness. This type of mind depends on language only as a form of crystallization between the medium of pure intelligence and its mental source.

The human mind may be viewed as a set of radiant drums which spin to form convection currents like a smoothly running engine. As each drum spins, patterns of heat/light wave energy create matter/thought which stretches out over the field of time for interpretation. Our present mental structure is symbolic. Our new mental structure will be literal. In other words, we will literally "think time" and it will be made manifest.

In this literal experience, the abstract value of mind is expressed more in terms of looking into the windows of psychology from a time continuum. We are able to look forward and behind ourselves, in a manner more deeply interior than we presently experience. The entire mechanics of our functioning, both voluntary and involuntary, is available to us. It is as if we are transparent to ourselves. When nothing is hidden, there is no need for a subconscious aspect of mind. Everything is out in the open and completely accessible.

The human mind is therefore involved in stretching the time variables so that beings on our planet can interrelate in more profound ways. We can then have in-depth collective experiences of consciousness at incredible levels of intensity. The type of intimacy presently available to us on a one-on-one basis becomes available to us through groups. At first these groups may be small, but eventually they will involve complex networking systems of many, many people.

As this collective value grows, the human being is able to spend a great deal of his/her energy developing a maximum interior relationship between the interfaces of his/her own consciousness. The time we spend in this dimension is engaged in understanding and mapping the most subtle levels of our own being. Each stratum of knowledge is explored in depth. When we investigate our awareness, pools of light open much like casting a pebble in the pond, and strata of energy/matter/consciousness are engaged. From this point, consciousness expands around the field of perception and the entire expanse of knowledge opens itself before us. There are no boundaries. Every point of knowledge has maximum resonance within our own field. We feel everything acutely, but strangely enough, are distanced from it through pure objectivity. Everything is very real but at the same time is experienced as simultaneous within ourselves.

The psychology of the future is involved in naming the codes that will create breakpoints in the different strata of consciousness. It will create shortcuts in the consciousness pathways so that different points of perception can become more clear. Each avenue of knowledge is specific in its infrastructure and requires different tools of perception. This will be the function of psychology.

Psychology also looks out for saturation in our nervous system. It is responsible for minding the physical body so that gates of intercession can be open when perception fills the nervous system to the point of overflowing. As new space is needed, psychology will develop consciousness tools for expanding, making room, consolidating. Psychology functions as a consciousness management system.

At the present time, psychology is the study of our limitations; in the future, it will find ways to handle the voluminous amount of energy that we will be learning to conduct. It is as if we are moving from a small chamber orchestra to a huge symphony of instruments.

Viewing the Body

Every system of the body has its own body psychology composed of the memory values of all of the functioning organs, their relationship to each other, and their hook-up in pure consciousness. The psychological value of each organ, its aggregate makeup from a cellular level, and its genetic imprinting and origination will be studied.

Every cell has its own peculiar psychology. It is completely individual. It is a small, living entity, equipped with its own habits and idiosyncrasies. Our own consciousness can speak to itself on a cellular level, enabling us to communicate at first with aggregates of cells, and finally to smaller and smaller levels, right down to the infinite. We become atomic consciousness, able to transmit frequency codes directly to the most minute aspects of our individual makeup. So, as we become more collective, we also become highly specialized, minutely individualized. Every cell is given economy lessons, and learns to function more wisely, more efficiently, more dynamically. Psychology is the science that teaches cells to think; they are taking lessons from the infinite.

The marks that consciousness makes on the cellular level create the speaking tools for the body to relate to the outside world. The environment that we live in interrelates with our awareness through cross-exchanges of subatomic structure. We literally swallow our environment and analyze it, homogenize it, make it our own. We become involved in teaching our cells the language of the environment. Each cell is given codes of conduct with respect to the individual character of relationship.

Our personal consciousness can work directly with the environment, and thereby change it. We are not limited to where we find ourselves. Through a knowledge of psychology, the body learns to make its environment. The body acts as an encapsulation of a temporary nature for consciousness. It studies consciousness, much like a living laboratory, and involves itself in developing creative outlets for cellular regeneration.

Each body can create its own environment and entertain itself with what it has found. In this sense, matter is infinitely expansive. However, if it is necessary for it to contract for a limited period of time, perhaps our equivalent of even a hundred years, it can do so voluntarily, with the intent to study, much like a controlled experiment in our present laboratories. Static focus will only be tolerated as a tool for discovery; it is not necessary to hold back anything once it wishes to expand.

The body ripens rather than ages. It develops spirals of interactive consciousness that tell it how to change the environment so that the cells can function more efficiently. The feedback loop between consciousness and matter becomes so refined as to make it very difficult to distinguish between what is solid or fluid. In this respect, we enter into a realm of fluid mechanics in which psychology functions as a language of economy for our ability to relate to the outer world.

When inner and outer are no longer so distinct, we develop new terms of expression for life itself. We become "biolingual" and are directed through impulse rather than through linguistic interchange as we presently know it. We no longer have to trap consciousness in order to interact with it. We are able to run along with it through the annals of time, keeping abreast of ourselves through the magic kingdom of our own understanding.

Transitions

To move from our present static state to one of complete fluidity requires a juggling act with respect to our relationship to spirit. What we know as spirit is the inner birthing mechanism for energy to circulate into matter. When we describe ourselves as "spirited" we are referring to our inherent dynamics with respect to activity. It is a form of enthusiasm, energy, and light.

Our present understanding of spirit is that it is something that makes us human, stirs our appetites, and makes us restless for new things. As we shift to a more fluid response apparatus, spirit becomes responsible for changing us over to a more frictionless and expanded view. We find ourselves looking for the smooth course through a given point of view and are repulsed by that which slows us down. As our psychology changes, we are attracted to possibilities that engage an intensity of spirit on a cellular level. Every instance of life affords us the opportunity to break up previous levels of understanding. If we welcome this salute to change and do not put on the brakes, we will be happy. Without the excitement gained by leaping over the back of change we will feel a tremendous sense of loss and grief.

Each of us is being challenged to remember our identity and to pull away from those constructs which perpetuate our perception of who and what we think we are. No definition of humanity can be complete at the present time, because most of the data is not available. We are not able to image ourselves as open-ended, because we are used to encapsulating our thinking in order to determine its merits. A human psychological system that looks at the open-endedness of things and seeks to develop the tiniest flappings of emotion within the heart, causes us to discover the most important part of the matrix of life, ourselves.

Psychology now has the opportunity to develop the science of origin with respect to pure mental activity. The mind can reframe its locator points so that everything refers back to a calm but dynamic base. From this center, the winds of specialization can develop. The body can redifferentiate itself, developing new systems, more subtle and more far-ranging than what we presently know. Psychology can study these systems and develop communications networks between them. These networks extend both telepathically and biomaterially throughout our own bodies and to those we are close to.

We live in a wraparound universe of form/energy/matter/thought. The membranes of the mind become completely transparent and the need for privacy as we know it is not felt. When every one can know every other one, then the trend of life will be towards developing the subtle breath that apprehends the wonder of the universe. We become creatures of exploration, interwovenness, passage-travelers guided by the lighthouse of time. Every person becomes a message unit for other travelers to lock onto and ponder.

No secrets can be kept; at the same time, no memories which are not made at the beginning are able to be realized. Every memory is an imprint of pure awareness, tied into the central nervous system, appreciated from every angle, then fed back out, magnified, cleaned and made anew. Every thought is a complex, creative web of consciousness developing foundations for new intelligence networks to expand.

Psychologists are those specialists who take particular relish in following the trends of consciousness to their root origins. They milk the value of knowledge and keep furnishing new avenues of possibility to those who wish to study a particular stratum of perception. Psychologists are the traffic cops of the human imagination, developing the routes over which further research can travel. Every route has its own system of checks and balances, stops and starts, ins and outs, the complexity of which is mapped from the inside out by the individual psychologist.

The Brain/Mind Field

In our study of human intelligence we need to learn to differentiate between the mind, which we may view as the substance produced through brain activity, and the governing energetic blueprints behind it. The mind, in its role as a circulatory pool for the field, is responsible for correlating thought, creating channels of activity for reason to take place.

The human mind is presently a gateway of response between the biochemical signaling of the nervous system and the energy centers at the highest range of the cranial field. The mind is the playing field for mental activity, but is differentiated from the range of motion of that field in and of itself. The mind switches intelligence from one portion of the brain to another through biochemical relays that create a synchronistic ordering of field intelligence in the human energy matrix. The brain feeds the mind restorative pathways and allows the mind to distinguish its holographic experience directly.

The character of the human mind functions through cues arrayed through the five senses; it is oriented more towards sight/feeling/sound than it is to smell. The animal mind is oriented towards hearing/smell; although it may create pictures, they are biorepresentational, rather than being arranged through feeling/emotion. We have the ability to change the pictures of our minds through direct knowledge and sensory output.

The biggest change appears to be in the range of motion that our present mental processes can direct, as well as the style in which this direction can come about. The present mental system is limited in terms of its ability to create causality between thought and action. We tend to exist in a shadow world of mental symbology, which coats the mind in thought constructs that are not functional in relative terms. In order to create synthesis in the mind, or concrete patterning, we are forced to rely on an interlay of rational synaptic experience which separates what we know as imagination from everyday functional concerns. Even for artistic people, the gap between what the mind can conceive and the body can render is great.

As we begin to evolve, we begin to have more ability to actually realize what the mental field identifies as desirable. The person becomes capable of painting the field of desire directly onto the canvas of life. The thinking-feeling mechanics become more uniform, more precise and easier to accommodate. Pure thought or pure reason is based on the understanding that thought and feeling are complete opposites only when they are not synchronized through the complexities of the human heart.

The sensitive biochemical union between heart and mind provides the opportunity for perfect synchronization between what the heart feels and the mind thinks. As the two become streamlined in their response, the individual is capable of lifting the palette of being from the usual interface of human activity and is able to "swallow the fish whole"; he/she now has the ability to create on sight. The individual is wholly present for all life events, developing a principal matrix of response between all of life's accomplishment and the fourth and fifth dimensional self.

The characteristic expansion into the fourth dimensional aspect of Self signals a return to being, an experience of the Self as the template for activity. There is a degree of unified awareness, based on the recognition that the environment is an outgrowth of interior perception. However, the doer is still recognized as something that exists outside of the Self, creating some separation.

In the fifth dimensional aspect of Self, the door completely closes on any perception of separateness. The Self is viewed as independent of all action, screened from any sense of extraneous detail. The mind, aligned completely with the unified perspective, cannot literally peep out of the doorway of itself without seeing all objects as part of the furniture. There is no relaxation of this experience, even in sleep, where the mind begins to recognize itself even in the fabric of dreams. This poised, alert, attentive state creates a sense that time is unalterable, unchangeable, while at the same time, ever-changing, unrecognizable from moment to moment. This lucidity of the timely state, this attention to detail even in the midst of absolute unification, creates a feeling of satisfaction that is without desire.

The fifth dimensional state allows the individual to travel to other dimensional vortices, where the time differential is of similar proportions. As the individual assumes the shape of time, he/she can return simultaneously to the houses of the past and the future without retaining the body as a lightning rod. The body remains grounded in being and can experience itself as internally mobile and essentially free. It can house the range of the infinite with graceful magnitude. There is no experience of differentiated lucidity because no experience can be perceived as perceptibly different. The engagement is on the level of consciousness itself and there is no dampening of this individualized yet collective matrix of experience. In the fifth state, the magnificent ease of the unified dimensional structure reveals itself as a never-ending map to the all-time.

When the individual can live in the storehouse of his/her perceptions unhindered by any deviation of response due to emotional eccentricity, he/she can experience a happier, more fulfilling and dynamic array of activity. Life becomes highly coordinated and purposeful.

The Nature of Mind

Mind is the principal element through which we probe into the mechanics of time and space. It is through the mind that the origins of our species can be assimilated. The human mind carries within it all of the root origins of its collective awareness, dating to prerecorded history and beyond. In other words, our minds know the area of life before we were human and can completely recapsulize our entire history instantaneously if necessary. The limitations set on our intelligence are simply placed there so as not to overwhelm the switchboards, until the proper mechanics can be set in place for the individual to wake up to his/her full potential.

As the seeds of origin sweep over the individual, one becomes aware of the nature of the array of potential interfaces in the field and develops an enthusiastic willingness and capability to explore them. All of the vast stretches of mind/field become open to one, and there is a great deal of joy in this expansion. The mind perceives its undercoat in the vast reaches of space-time, and it sets out like a skilled explorer, navigating its options.

The mind that is released to itself becomes essentially free. It can develop chains of relationship between the different pathways of pure knowledge and can learn to link itself in these fields as need or simply curiosity mandates. Once the mind has realized itself to be infinite and without fear, it can leave behind its limitations even as the shadows of such limitations seem to be still in place. In the transition, this is an important consideration, because although our minds are still filled with the illusions of limitation, we can burst open the seeds of reality without fear. Once we understand how to focus our intelligence, the mind can go more deeply into the strata of wakefulness and search out our priorities. This is the beauty of an open mind.

The Qualities of Arrayed Perception

Each individual has a complex system of overlapping matrical fields which constitute the brains of his/her reality. These fields, when functioning properly, are synergistic in nature. Each layer catalyzes and jump-starts the other, developing a chain of command that allows strangers to become friends with regard to the uptake of new or difficult information. The mind wraps itself around new or foreign material and literally swallows it, allowing it to be assimilated by the entire energy system.

The mind is a spiritual messenger for the imprints of what we would term the soul. This core intelligence, which is the absolute value of our existence, is reinterpreted and assimilated through the brain/mind field in a system that provides entertainment for our human selves. We are the players in this game and we create all of the pieces that go down on the board. As we learn how and where structures of consciousness arise, we are able to create fields of response that better promote creative activity. This interactive process, which is response-neutral with respect to our environment, is at the same time completely contingent upon it. This is the wonder we call life itself. The mind plays tricks on the heart in order for us to learn; the heart plays tricks on the mind, to keep it from overpowering our sense of proportion.

Since our environment is not static, but is instead constantly in motion, our senses must constantly be lying in wait to create new fields of perception. These arrays of perception become more unified and coherent as our evolution progresses. It is as if all of the colors of our intellect are at first scattered to the winds, but as we develop, they become more orderly, careful, and precise. Our entire range of feeling/thinking/motion enters into a state that can be described as creaseless, without folds, open to the mediating influence of time.

Our environment is completely time-dominant; everything we see or feel in our world is based on the timeliness of it. It is living its presentness in our awareness, because that is where we are couched now. If the environment were to speed up, slow down, or essentially alter its time values in any way, the mind would have to adjust to match it.

The mind recalibrates itself countless times each interval in order to compensate for drops in the time value. What we see is not a nonmoving, nonchanging value, but one that is constantly moving, constantly changing. In order to compensate, and give us a feeling of security and order, our minds have developed relay systems that keep us in balance, allowing us to feel like we can motivate our destiny through our environment. This gives us the quality we would term "ego."

The ego is developed to enable us to feel that we are at the center of our personal and collective universe. Without this, a sense of collapsing into the underbrook of time would overtake us. We would not know where we began and the rest of our world left off.

As our response becomes more fluid, we deny time its value, allowing it to shake up the accustomed field of response. This may be outpictured as a series of retakes in relation to our personal environment. That which we have known as true, fixed, or retainable must be made anew. We are causing our situation to loosen up, change shape, change character. Without a system to loosen the armor of reality, our internal evolution would have no room to show itself on the outside. The arrays of possibility would be extremely limited.

Loosening the Screws of Our Environment

As we retake our field of perception, we are capable of restoring the dimensions of our true selves, entering into a time/field gate with our immediate environment. That which we have distinguished as ours becomes capable of entering a portal of perception in which we identify the personal as simply part of the all.

We embrace our possessions, friends, centers of activity, as part of a fluid, interdynamic chain of being in which we appear to be in the center, but which we simply realize to be part of the all-that-moves. As the blocks of time appear to reestablish themselves, we entertain the possibility that we can step back, away from the immediate perceptive response of our environment, and look out upon it. We enter into a laughing union with it, allowing us to restore order to areas of our lives in which disorder had been created.

The orderliness of this perception is inherent in the degree of changeability present within it. We are allowed to remake the co-interval of expression between our nonchanging and changing voices. The interplay between unity and divinity becomes so effective that we can hold our own, feeling free to meet our destiny in relation to union with infinite perception.

Our environment becomes a fantasyland of pure interimaginative realization.

Although we are still locked into the collective constructs of those fields of endeavor we find around us, our ability to play off them and interchange the array of moveable parts becomes more and more distinct. This creates a tremendous surge of joy and happiness in the individual, and allows for new breakfronts in consciousness to be established. Tunneling through the disorder we have created may at first seem to be an insurmountable task, but once we have lifted free just a little, the glimpse of the possible becomes a pronounced vision in our own perception. This creates the mental material for a new mind. It is an event of both wonder and responsibility.

Changing the Moveable Parts

The circumstances in which our life's destiny has placed itself appears, at first, to be nonmoveable. We appear to be random figures in this interplay, without power or possibility of compromise. This notion of powerlessness is easily extinguished once the mind becomes capable of interplaying with the field in such a manner that it can open new variables in the response mechanisms.

Once the mind can identify itself as the originator of reality, it can develop new frameworks through which it can promote its chain of command. Unconscious perceptions become ever more conscious, and the mind identifies itself as the chain of command for pure being. From this seat, we can view ourselves in a new way. We are able to develop new enchainments in the imprints of consciousness; it becomes the renderer of its own interface with time.

Consciousness looks out onto the window of time, develops an interactive relationship with it, and models this relationship for the mind. The mind takes up the lessons of consciousness and develops practical applications for them. The range of creative response is greatly expanded.

Once the limitations on valuation are lifted from consciousness, the mind enters a zone of "no time" in which it can coast along the surface of its constructs and reevaluate their importance in light of present circumstances. In this way, the mind literally reconstructs time. It develops interlocking units of behavior that stretch the intelligence and allow the individual to become free of past constraints.

This sense of internal freedom can be intoxicating. It is up to the framework of absolute being to develop a system of holding back the mind until the full measure of careful response can be invoked. Once we are free to unleash our minds, we must continue to dip down into the field of pure response, that absolute knowing, for a sense of true and careful mediation to occur. We must lubricate the joints with pure consciousness so that the entire sensibility can be realized in an organized and uplifting way.

This state of evolution allows the parts of our lives to become more moveable, more flexible, more easy to change in shape, color or form. We are playing on the field of the infinite with our own understanding and this understanding is no longer shackled by an intellect that is fearful of change. The entire structure is more easily retrainable into the mechanisms of time and knowledge.

Systems of Therapy

When people seek assistance of a psychological nature, they desire to leave behind certain preconceived patterns of thought and behavior that inhibit their functioning. Precluding any intensive physiological disturbance, they are seeking to retrain the consciousness so that more refined fields of perception can be realized. Psychological problems are really manifestations of a consciousness that has not ordered itself with respect to its ability to view responses effectively. When responses cannot be ascertained, the mind starts to fumble around in its own territory, seeking mind-cures for what essentially must be corrected on the level of consciousness.

This attempt to correct the flow of consciousness through the mind is ineffective except in assisting consciousness to seek clues with which to repattern. "Talking therapy" can open doors to the consciousness by identifying areas of difficulty, but it is consciousness itself that must pierce the boundaries to open the gates of realization so that the new inner territory of knowledge can be gained.

Consciousness, by its very nature, wishes to flow out onto the playing field of perception. Pure consciousness in its free state will spread to every corner of perception, opening the possibility for absorption into clarity without having to search the mind. This is why it is so important to practice those disciplines that will expand pure consciousness and allow it to lubricate the mind in order for revelation to occur.

The advanced concepts that are formed through the interplay of pure consciousness and mental activity are themselves the basis of therapy. As the mind grows into itself by developing points of entry into its own field, it can develop systems of analysis that bridge the gap between what we might term the mind and the heart. Emotions in their evolution must go through a continual process of refinement; so must the mind.

The mind enters into a state in which conceptual knowledge can spring forth, apparently effortlessly, from the constructs of consciousness. In this state, every problem has its solution. Even if detailed access to understanding is not instantly forthcoming, it can be placed on a level of energetic activity in which it can simmer until the proper response is realized. In this state there is no unknown and therefore no confusion. Confusion is not possible because the knower can be consistently and easily called into place at every turn. When the knower operates from this level, a sense of doubt is stricken from our awareness. Everything that we choose to understand we can understand.

Problems in this transitional state are simply matters of lack of will, gaps between what we already know and do not wish to know. The function of denial serves the transitional makeup, so that we will not be overwhelmed by sensory input that perhaps we are not prepared to assimilate. As the physical, emotional and mental vehicles are prepared, we are able to assimilate vast quantities of information, analysis and understanding.

As the science of psychology begins to understand the operational mechanics of consciousness, it will be able to develop therapies that open the fields, restore order, and create energetic links between old and new strata of awareness. The psychologist cannot draw maps of the mind until he/she is familiar with the territory that is to be mapped. The lay of the land lies in the territory of consciousness within which different subtle strata materialize through human feeling and perception. The psychologist must follow the lights of perception and learn from them.

Forming Vibratory Links in Consciousness

The person who wishes to open the gates of perception must be willing to first link with his/her own source of pure awareness. Once this link is successfully achieved through meditation, then the individual can become capable of unifying the source of his/her intelligence with the complexities of the fields.

This primary link may sometimes be facilitated by bathing in the field of one who has firmly and successfully established this link in her/himself. Sometimes it may also happen through an identity crisis brought about by severe physical trauma, accident, or loss. The reason for this is that at these stressful moments, the individual is broken apart from his/her conventional identity stream and is forced to enter the realm of the infinite in which he/she is sourced from the Host level. Once this occurs on a firm and conscious basis, it is impossible for the individual to go back to a completely unconscious state. One glimpse of Self is sufficient to keep the individual searching for more. The craving for the infinite is itself infinite.

Once the primary link has been established, either in the short term, or in a more viable, reclaimable way, the psychologist can enter into a state of merger or union with the individual. In this process, he/she is led back to this source place at every session. Most of the therapist's job then takes place on an energetic level, rather than through physical or verbal contact. The individual becomes linked with the therapist in a therapeutic symbiotic union until such time as the client is rebirthed onto the field of the Self and can function more independently.

This experience is a subtle one, but it is also intense in that the individual may experience a profound degree of devotion to or dependence upon the therapist for a fix on reality until the client's reality becomes clearer. This transference is a necessary energetic phenomenon, unless the initial transformation occurs through a completely stabilized and perfect transmission. This can be achieved through union with a very advanced soul. For most individuals, however, the fixing of the infinite occurs incrementally, according to the degree to which the will is capable of letting go. It is a matter of time, patience, and willingness to change.

The energetic field develops linkages at different steps of matter/time ingenuity and in this manner leaps back and forth between clock time and absolute time until it can transfer itself directly into the one-pointed field. As the identity construct is redefined through therapeutic progress, the individual becomes capable of linking with the codes that produce the matter/time framework itself.

The therapeutic process cannot reach full stabilization, therefore, until the individual is literally able to make matter with his/her mind and thereby become the source of his/her creation. Until this point, the individual is like a child dependent on the environment to give sustenance and unable to develop a completely equal partnership with it. Until thought can become matter through proper training, intention and alignment, it is simply ephemeral rather than concrete, literal, and powerful.

The individual must develop the strength of consciousness to link directly with the source of desire and make it manifest. It is the job of the therapist to provide a model for the individual of the link between consciousness and matter. The therapist then helps to restore those links and allows the individual to imitate the constructs until they can be more perfectly his/her own.

Modalities of Therapy

New therapies which prove most effective will speak directly to consciousness, and will therefore be energetic in nature. By calling on the terms of consciousness, they will allow the individual to speak fully with his/her inner self and develop pathways to link directly to physical, emotional and mental systems. The individual will learn to tunnel his/her way inside and experience the entire field of physiological, emotional and mental functioning.

The person must make a deep study of him/herself. Every therapy session should in essence be a fantastic voyage into the mechanisms of consciousness. It should provide a passage for the seeker.

The properties that will restore consciousness are within the seeker; however certain environments can be created which make this retrieval process more conducive. These include therapies that involve elevation of the five senses to more pronounced levels. For example, therapies that create experiences of advanced touch, hearing, smell, or taste allow the individual to look into the self through different lenses of perception.

All of these levels must be accompanied by a greater degree of inner visual perception. The visual field underlies all of the other levels in that the individual must see him/herself bare and unaltered in order to perceive what he/she truly is. When an individual is blind to the source of pure radiant energy, or light as we might call it, the individual is blind to the nature of being itself.

The visual medium is the symbolic representation of the Universal Light, which is the causality of the Universe. Even when the individual is physically blind, such perception can be instantly available once recognized. When the individual can see, he/she will more readily be able to hear, taste, smell, or touch. The pathways to the other senses must be entered by seeing the routes.

This type of sight is made available through the individual's developing a living knowledge of his/her own inner and outer space. One must be able to sense the coloration, continuity, and formation of the environment and to reference it spatially in one's own thought/response mechanisms.

When the person views the outer reality as constant feedback to the inner, and the inner as a creation matrix for the outer, he/she wakes up to the causal plane of his/her own field and develops the power to transceive wave frequencies of the infinite. He/she is no longer lost to the creation of waves in his/her own cosmology. This allows one to change the script, the mechanisms of response formation, and thereby change the constructs completely. The first few times this capability is realized are the most difficult, because the person has not learned to trust his/her perception. Once the clouds of difficulty have been cast off, the individual can look at the world as a nonlinear, non-time-based construct.

The most effective therapies, therefore, are those which can change the perception of time and alter the degree of functioning between unilineal and causal fields. The ingestion of natural plant substances into the body to accomplish this can be effective in precipitating this change; however, it is up to the consciousness to hold its own with respect to these changes.

The therapist who can evoke vibrational points of reference that address the individual's lack of perception will more effectively create seeds of change. The individual must be allowed to see and hear his/her own unclouded mental reflections and to enter into a state of clarity in which the shades of reason are more pronounced. Verbalization can crystallize this clarity, but only when it is coupled with a mental-vibrational linkage that can direct consciousness along more advantageous pathways.

When the nervous system can develop a printed language of feeling between itself and the fields of intelligence that magnify the heart, the individual can develop a full range of motion between that which he/she feels and that which he/she knows. This differentiation between knowing and feeling must become so keen as to be automatic. The stronger emotions such as anger, sadness, hopelessness, must be tempered, and their place in the array of feeling must be understood. Until the individual can look into the source of a given feeling and have the courage as well as the know-how to change its range of motion when that feeling is inappropriate, he/she will not be able to enter the full expanse of higher mental faculties which are his/her birthright.

Intelligence must be based on a free flow of feeling that is tempered, even and uniform. To accomplish this, however, sometimes an inappropriate or exaggerated stretch of territory in relationship to the feeling level must be uncovered. This is seen however, as a temporary solution rather than a permanent resting place in which consciousness resides.

The person who has cultured consciousness most clearly can perceive the emotions, even in their most tumultuous form, as a source of perturbation in his/her awareness. They are like the weather of the heart. However, like our response to the physical attributes of weather, we would not proceed to go out and conquer the thunderstorm but rather view it as an inevitable part of nature's organization. Until the individual can meet these changes with a degree of equanimity, neutrality, and calm, he/she will not be able to enter into what lies beyond them.

The degree to which one understands the source of these stormy seas in one's own heart determines the extent to which one can find a way past them. Since the transitional consciousness still perceives time in a timely state, he/she seeks to uncover the message units of truth that lie in past events. In this way, the individual sets up constructs which cause the seeking out of the origins of primary emotion. These constructs do indeed lie within the individual matter/energy matrix and can be sought out, but it is really the flood of pure consciousness that ultimately corrects them.

By understanding problems at their source, the individual can more profoundly discern the effects of neglect, abuse, or lack of compassion, whether directed at oneself or towards others. Without turning a blind eye towards such cruelty, the individual can realize the underlying source of injury, supporting the willingness to forgive. Such intense hurt can cause the individual to seek to leave physical form. Once the individual becomes completely free of the desire to leave this life based on unresolved pain or need, he/she will no longer be prone to leave before the soul's plan is finished. He/she can then unite with consciousness so that it can create optimum benefit.

Thus, the job of the therapist is to provide support for crossing the ocean of despair that envelops the individual makeup while it is still unable to recognize itself as immortal and infinite. Once this recognition is gained, even in its incipient form, then the totality of healing can occur.

Breaks with Unity

The individual psychology creates an approach to life that provides a complete panorama for consciousness to steer the way for the mind. The mind functions as an effective tool in actualizing consciousness, once the individual realizes the full potential of his/her thoughts. The goal is to steer every thought in the direction of unity of origination, whether leading into emotional or conceptual territory. If this is accomplished, the individual can then begin to experience a reunification of perception that can be truly startling.

Our experience of thought is constantly separative in its common functioning; i.e., we use thought as a medium to distance ourselves from points of origin in consciousness rather than bringing ourselves closer to it. The new modes of therapy will constantly seek to bring the individual back to the antecedent thought/resonance pattern in consciousness until this process becomes so habitual that he/she can unite more directly with Self.

An example of this might be the thought, "I am lonely." Once the individual views the thought, identifying it as separative, then he/she can be led back to an awakened understanding that the source of connection is profound and complete. Even where apparent emotional union cannot or will not be found, the individual can return back and re-experience the freedom of such consolidation.

This is more difficult to grasp, perhaps, when the thought processes are more conceptual or intellectual in origin. In the thought, "I do not agree with this person's ideas," or "I feel that my ideas are more conclusive," the separative nature immediately makes close distinctions between the "I am" reality and the "I am not" reality. We simply view ourselves as disconnected from the sameness principle inherent in every interaction. When the individual starts to strive towards sameness in him/herself, and begins to view reality as a process of symbiosis rather than constant differentiation and distinction, a shift occurs in consciousness and the person is able to ride the ups and downs of life in a more productive way.

When every individual circumstance can be viewed as part of Self, then it becomes possible to actually see the true differences. These differences, which are subtle expressions of consciousness, cannot be understood while a more coarse or disjunctive voice is consistently present. This function of intelligence, to discriminate at the expense of reason, is the main stumbling block to greater experience of unity on a psychological level.

Consciousness contains a powerful drive towards totality; it wishes to view things as part of an integral whole. Splitting ourselves into pieces (through analysis, promulgation of opinion or ideas, or emotion that pushes away the possibility of reception) can block the way for pure consciousness to smooth things for us. This limits our point of view and decreases the possibility of happiness.

This shift into an intelligent sameness with respect to identity is most crucial to us. Once breaks in unity are seen as something to be avoided, an experience of unity will be more easily reached. Laughter at the expense of separateness appears to be the key ingredient to this. When the individual can view pain as a state of dormancy of spirit, rather than cherishing it as a profound reality of presence, then the pain can easily be transcended.

The notion that pain must be cultivated in order to be fully understood is a common psychological principle of our time; it is relatively new, however, in the scheme of psychological reasoning. When pain can be viewed for itself, yet rerouted to a different point of origin, it need not be simply suppressed; it can be truly brought back to its source for reevaluation. Consciousness is the true measure of relational pain. It is the best judge of its merits for continuance. When the individual routes the pain back into pure consciousness, the system will on all levels judge the merits of the action. If the pain is indeed necessary as part of the natural order, it will necessarily and absolutely continue. This may be attested to by those healers who commonly seek to eradicate pain in its various guises but find that it springs back when it is not ready to be released.

The notion that pain is in and of itself a bad or undesirable experience is the source of this problem. Once pain is recognized as true friend, which it will be when it is not chronic, unconquerable, or separate from awareness, it will be viewed as a productive leader in sharing its message with us. Chronic pain of a psychological or spiritual nature is often unnecessary in this regard, and its continual formation is simply brought about by lack of experience in knowing how to alleviate it. This type of ignorance is in itself painful. It may be our greatest pain of all.

Symbiosis in Consciousness

The human being is functionally a team player with respect to his/her development. One comes to Earth for the benefit of relating to other human beings and sorting out the truth of one's own intelligence. All of the work that we do, the people we respond to, the insights we maintain, are related directly to our ability to live inside the skin of other individuals and develop compassion. There is no work separate from this, nor more important than it. Even the most complex intellectual development pales when compared to the development of an operative mechanics of mind/feeling in tune with the environment.

Life on Earth is by nature motivated by sensory output. Life in other dimensions can be more efficiently creative in some respects because we are able to access pure intelligence more directly, without having to share it through the skin of our physical bodies. Conceptual understanding can be gained instantaneously once the true heart has been realized, but without the heart's full development, there can be no true intelligence. We are placed here on Earth, therefore, to discover where our hearts are bleeding and sew them up through the balm of expanded awareness.

Spiritual teachers who place emphasis on the development of compassion know that right reason cannot be established until every shred of mental activity is encompassed by an understanding of its cause and effect. Without this we are like children who realize ambitions without understanding their lasting effects on those around us. This is, of course, the present state of affairs in most respects in relationship to our collective value of reality.

It is therefore of the utmost importance to renew our vows to the infinite by unmasking our feeling/response mechanisms, viewing our enemy as ourselves. Every small nick on the skin of our relations must become a relational pullback to our own inner reality. At the same time, we cannot become bruised by those stresses and strains that seek to limit the creative wellspring of consciousness. Left to our inner dimensions of feeling, clear dimensional boundaries must be established. These boundaries are based on moving the core of being inward so that the outer crossings are made without having to step on others' toes. Each singular experience can be seen as a reliance on the ground of form in order to build substance.

When we leave this Earth, we are left with a matrix of awareness that builds interrelationship more deeply into our vital mechanics. We are seeded with dreams that are the cornerstone of our next reality. Every curve we take, every understanding realized, prepares the way for a fuller dimensional swing once the physical body is shed. Even those of us who will reach perfect immortality in this physical dimension will continue to develop curves of expansion into relational experience.

As long as we are human we will seek the Divine, and perhaps it is the secret of the Divine that it relates to itself through the living experience of its own creation. It is both infinite, all-knowing, and yet ever-seeking, ever expanding. This is the nature of Intelligence itself.

Meeting the God Structures

Our experience of God is based on our ability to witness the development of consciousness. The personification of God as a namable, definable presence by its very nature limits this view. Once we understand that God is not understandable, in a sense not attainable, we are given the opportunity to attain. We witness this unreachable place through striving towards our own unification, relying on the strength of wisdom teachings passed down to us by our cultural and religious heritage.

If history is acknowledged as an aberration of time and is in a sense ahistoric with respect to a completely linear sweep of reality, then God must be viewed as an ever-expanding, timeless and non-distinct Presence. Our view of history has frozen our perception of God into ancient terms that are not in keeping with His living language, his revealed memory. As Mother/Father God reveals itself, the pattern of events we term as history can be swept forward; we are no longer fixed to a view of ourselves as a sum of the events that shaped us. Even the events themselves may be altered, reworked through the time-traveled doorway of cognition until we view such events as lucidly capable of alteration, even as we can alter the patterns of our dreams.

The God that lives inside each of us is our own, but there is a wider God, a more expansive sweep of consciousness, that shares its breadth of being with every apparent and nonapparent visitation in the Cosmos. The flowering of God, through the underlying matrix of expression, is the great mystery we encounter as we study the mechanics of Universal feeling.

As we depersonify our own experience and rely more heavily on our inner namelessness to find our moorings, God itself can be stretched to more infinite proportions. That which is unnamable is that which we are. We are the Divine, nameless expanse of pure energy, pure intelligence, that gives life to the lifeless. When God remains comfortably unstructured in our awareness, we are allowed to lift the lid on His/Her radiant reception and enjoy the pure bliss of feeling/nature, which is synonymous with the experience of knowing.

As the psychology drifts into a sea of timelessness, it becomes operative from this nondistinct source and can identify itself more as an avenue of changing feeling than as a fixed array of parameters. Without opinions, judgments, perceptions, or nuances of personality, we would appear to be hollow or empty. The surprise is that as we lose all of these qualities and are freed to our own devices of intelligence, the inner light laps up the merits of our actions and grows more profound in its appreciation of life itself.

The distance that we cross into the dimensions of being is so acute, so awe-inspiring, that the measure of ourselves proves to be more of what we knew of God than of anything else. We are taking a rod to the Infinite and do not find ourselves wanting. We are all or more than we ever dreamed of or could possibly imagine.

Breathing Thought

When God is realized as a sitting value in the mantrum of stored thought, thought itself can become an avenue whereby God is realized. Every thought creates an outpouring of appreciation for the Divine. It is a poem to the Infinite. No longer able to develop separative thoughts, our thought becomes instantly magnified. We are restoring granite to our actions.

Thoughts such as these are living memories of God. They are monuments to His/Her breath, His/Her sensibility of feeling. When thought becomes elevated to this position, it is not limited by any previous relational value. It is wholly and entirely its own, and therefore can match itself to the memory of God. This is the highest function of memory. It is restorative. It lifts us to the Divine plane on which we are able to remember those activities, or points of doing, as part of the Host Creation. We are mimicking our essential selves through activity.

This range of experience then becomes a crystallization for God Consciousness in our activities. We are able to live life from a purposeful and productive place in which we define action as stirring the waters of bliss. Without harboring prejudice with response to activity, we can view each day as a search for the

Divine Light in every creature. This is not purely figurative; it is an expansive, living reality of motion, pouring through every breath. We are breathing the skin of God; as it enters our smaller, more limited view, it stretches outward in every direction. Our intent is to live in the eyes of God; our vision is brought closer to the desired reality of wholeness and clarity. The most beautiful thing about this is that every small step we take in this direction is to be coveted and cherished. The wink of an eye we give to a stranger, the love we feel for someone who has passed away or who has touched us; these stored memories are re-realized until they become the schema of activity of our conscious and unconscious mind.

When the mind is free from thinking of itself as the all-knowing purveyor of force, it can become a light on the bridge, sharing itself continuously with Intelligence until consciousness can be experienced as truly uniform. In this sense, every inner attribute is posted on the calendar, evaluated as to time, date, and place, and recalibrated in future/present/past parameters. We are living backward and forward time breaks and storing reality for our own pleasure. This is the greatest pleasure that God gives us, that of hope.

The Language of God

Our experience of language is based on our feeling/perception derivatives. We are tied to the speech patterns that we elicit because they are the foundation of our travels in reality. We measure ourselves through the intake of language; we share ourselves through altering the perceptions of this living language. Although we generally understand that these constructs are painted in time for our benefit, we may not always realize how limited in perception they actually are.

The language through which we are able to perceive the Divine offers us the possibility of fuller understanding. Every Divine precept is painted into the doorway of perception via the intake of pure consciousness. We open the door to understanding through the unmasking of limitation. Each derivative can be secured through our own breath, our own usage from a purely physical standpoint. The language that we seek is the language that we are. We are the carriers of pure linguistic forms; we distribute these forms throughout our being and into our present environment. The storehouse of information and memory that such linguistic forms carry is the same storehouse by which we judge or view our present cultural understanding.

Pure language is the study of impulse breaks that occur on a subtle level, traveling the channels of speech into spoken form. The brain/mind functions as an interpreter, seeking to expand the range of what is possible, what is infinite. We are asked to function as tool-carriers for an intensive climb to the unsung regions of the Universe. In order to facilitate this climb, our entire matrical range of speech/tongue perception must be altered. We are being asked to develop language that is at once subtle and refined, yet at the same time implicitly direct and straightforward.

In order to accomplish this, our present use of sound must be extended to hear the subtleties of breath resonant in our speech patterns. The expressed linguistic stopgaps in consciousness are signals from the brain/mind to pure consciousness to begin the sequences that create spoken language. When language is a living expression of telepathy, it is not necessary for it to be as fully sequenced nor brought so fully forward as it is now.

Ancient languages were closer to their linguistical stopgaps because they did not call for such complex elaboration. Much of what was spoken was heard inwardly before it was expressed outwardly. The outward expression touched up what was already known. Our present need to put it all on the table has caused us to develop complex systems of wave formation important for our scientific and emotional development. It may be time, however, to switch back to language forms which are more thought/feeling derivative, closer to pure sound rather than fully developed inclinations of speech.

The tones and pulses of language speak directly to the heart. They are the outpourings of the soul's own need to understand, to express itself on the level of feeling. The core essence of Self speaks in the language of whispered revelation, calling out to its origins. We must develop these subtle, refined messengers of perception in order to speak to our inner being about the wonders we see there. Our inability to use language as a means of expanding consciousness can cause us to keep impounding consciousness in a relatively limited framework. This is why a new language for God is now so appropriate.

When we listen to music, especially music that is pleasant rather than jarring to the ear, we begin to imagine what language would be like if it was composed to complement the spirit. Language forms evolve much more slowly than musical forms; this is unfortunate because language is closer to home for most of us than the music we listen to for enjoyment. If language were treated in such a way that it was an act of creation at each and every moment, always novel, always original, we might not feel it would be useable because it wouldn't be standardized. However, it is this standardization that has robbed us of our understanding of what human sound actually is. An ideal experiment for those of us interested in refining consciousness might be to spend time working with sound in its pure form—to compose symphonies of communication that would be designed to develop certain metaphors of consciousness and enter them into the collective field.

Music often takes the form of instrumental expression in our culture and this is helpful, but pure sound, developed by the human tongue, might ultimately be the most powerful form of living symphonic expression. The range of sound, the quality of its integrity within our own brain/mind boundaries, interprets consciousness for us, and allows it to sing sweetly to the ear. Language that can express the matrical interlinks between pure consciousness and form can generate the signal impulses necessary for matter/creation.

In this sense, our electronic machinery, particularly computers, may be onto something. When we allow the machine to talk to itself through the beeps and whistles of electronic generation, we are mimicking the wave impulses that form the stretch to our own language. What we have seen through such experimentation is that memory responds very nicely to this system. Our abilities, however, are not matrilinear, as would be a computer or electronic device. They are sound-dimensional and unlimited by space-time considerations. The chorus of pure sound that can emanate from our own vocal apparatus could be so powerful as to literally stretch to other dimensions of inhabited galactic experience. Tonalities which make this mark are interior buzzwords for a new type of spoken/written vocalization.

What we must do is what a baby must do. We must mimic certain sounds. However, these sounds are not derived from the outside but from the inside. We are being asked to listen to ourselves so acutely as to hear the working of the kidneys, the heart, the spleen. We are being asked to imitate and thereby connect to our own physiological chambers of resonance. As we begin to accomplish this, the patterns of tonality of our own body/mind become available to us.

Since every being is entirely individual, its ability to speak is individual as well. Collective language has served us because in our culture, without such a tool, we would be unable to share our perceptions. As we evolve to a more telepathic or *oceodynamic* framework, we will be able to live language in such a way that it need not be entirely collective.

It is then that language can become a more well-tailored and original expression of our inner experience of space-time/motion. We will literally breathe language, and the sounds uttered will be distinguishable because they will be led directly to the feedback-loop of pure consciousness for interpretation. Everything will be known to us because there will be no differentiation from that which is.

The silence kept now by great monks or sages is storing away memory for us till such a time when we will be able to hear the literal sounds of silence behind that which now appears absent or vacant. Linguistic imprints will be able to create memories that are tangible, tasteable, and feelable, by all five senses–memories that have not as yet been categorized as sensory. Every syllable will be a treat for the tongue. and will relate to every aspect of life. This is the language of God. It is a full-ranging, synesthetic effect of consciousness, a mouthpiece for the Infinite that ranges in all directions of perception. It is a unified awareness that is all-knowing in its comprehension, yet simple in its translation.

With our new linguistic abilities, we will be able to translate the memories, perceptions, and feelings we now have into cross-translational bridges with other organisms and life forms on our planet and beyond. Our encoder mechanisms will be switched on so that there will be no gaps in our ability to send or to receive. As we translate our own living breath into language, into sound-sending capabilities, we will be able to receive the many channels of sound that are presently available for us. The stations of consciousness will open to us, and we will be able to find out the news in our own backyard.

Living Memory

Language that speaks directly to the heart, bypassing what we presently know as intellect, is not unintellectual in nature. It is still quite complex and quite capable of expressing what we presently term as concepts.

This new form of language is going to be highly visual as well as auditory; it will, in a sense, be like television as compared to radio. We will have more capability of holding memory when it is a causative value, because we will be able to call it up so distinctly in our frame of reference. With the shadows of language gone, we are capable of transceiving language more directly and vividly.

This extends beyond the realm of imagination to pure causation of consciousness in manifest form. Our sound, our breath, can actually make matter; matter, as we know it, is simply a repository for energy/consciousness. This magical quality of language, which has been known throughout the ages, will not be a mystical value relegated to a few, but will be available to all. When language functions from this memory, God can imprint His/Her wisdom directly upon us. We ourselves will become living memories of that Universal Will, that Universal Desire. Our psychological boundaries will be broken, and we will be able to recall at will that which we are.

With our memory restored, language will be able to define the beautiful parables of infinite realization that can be available to us. We will be able to call our names and immediately know what we are to be. This will limit all manner of confusion and suffering we presently experience.

Functioning in the Transition

As we begin to experience the linguistic breakup of our perception, the forms that we hold no longer seem functional. This could create a great deal of fear. It is our responsibility therefore, as we mature in consciousness, to welcome this breakup and to develop transitional forms of communication as well as matter itself. In this way, we can usher in the full change, trying it on before it is fully ready to occur.

Although the temptation is to figure out what will occur and to move quickly to realize it, this is really impossible, because the forms we take now will not last very long. When we view everything inside and outside ourselves as transitional except for that absolute center of being, then we can witness time changing its metaphors without our seeking to cling to it for dear life. Our definition of life must become centered on our willingness to explore consciousness from an upward slope on the domain of pure attention. If we can offer help to those who are lost to themselves during this time, or who have gone so far and so fast that they have no anchor, we can serve the Universal Will quite well.

Though the full transition may take many hundreds of years to complete, time can stand still for us if we allow it to. Our new psychology must be one that does not allow time to bind us, but instead makes garments of infinite perception for us to try on and to discard as necessary. Everything must be disposable because the very rules by which we define reality are changing. When what is up is seen as down and every facet of perceived conflict is torn away, we are simply left to interpret the principles of unification. Our route into the circle of reason within our own consciousness is the fabric with which we are left.

Transitional times are always difficult because by definition they cause us to wake up to something before that something is known. By valuing our humanity as intangible, not fully expressed, but available, we can enter into a realm in which possibility is always present and change inevitable. This is the means by which we can make the transition pleasant and creative rather than frightening. A breakaway psychology is what is needed now.

CHAPTER THREE

THE COMPLEMENTARY BODY

CHAPTER THREE

THE COMPLEMENTARY BODY

Lifeseeds

The human body is the outgrowth of universal evolutionary principles circling back onto themselves to produce miracles of structure and function. Every cell of the body is full of the life stuff that is spread throughout the Universe. Our bodies represent the primary seeding mechanisms, the life enhancers which develop the coat of invincibility that shapes our genetic blueprint. The relationship between the kingdoms of light that radiate throughout our cosmos and our own internal form is complementary in nature; it speaks directly to the metaphors of proximal creation.

To know how we are made and what we can become demands a knowledge of our full evolutionary structure. The genetic replay of our own *God structure* glues together that which we are; the constructs of our imagination provide the seeds for infinite realization. As our sense of purpose and animation becomes unlimited, so do our bodies. This is the secret to biological unification.

Every genetic strand has within it a coding mechanism for the full flowering of our intercedent matrices of light, heat, and energy. Even the chromosomal deficits that are perceived can be easily translated into new coats of armor via genetic transfer, both organic and mechanical. Everything about us is essentially replaceable, remakable, and reusable; there are no throwaway parts. Once we understand the complete transferability of our human body, we can begin to appreciate the tremendous gift of life that it represents.

Entering the Vision

We presently see humanity as a differentiated mass of people, racially specialized yet united in some inimitable way. This specialization along racial lines holds the promise of our full development. In each racial strand lies a germinal imprint of consciousness, with characteristics that are being stored for our primary development. Those species which have lived on Earth in the past are part of our genetic marketplace of stored information and possibility. The blueprints of all species can be reaccessed as we become aware of how the genetic monopoly can be cracked. We are the superstructures of our own imagination, and as such, all of the players that have come to our table provide the message units from which we can work. There are no extinct species in this sense, only characteristics and memories which can be recalled as we learn how to dive into pure knowledge.

Our mastery of creation must be based on our records of infinity, but these records cannot be made available until we comprehend them. From our present vantage point, we sit on the brink of the stars and wait until our fundamental consciousness aligns itself more fully with the Universal Will. We must regard ourselves as explorers in the field of mind who have been given the opportunity to venture into the footprints of creation and experiment with our individuality. In this sense, nothing that proceeds from our own awareness is frightening or unmanageable.

As consciousness develops it must create the technology that will allow full expression. When pure consciousness is extended completely into the field of reason, the outcome can only be reasonable. Pure consciousness has no motivation except its own infinite realization; therefore, it can be implicitly trusted. Those who open the gates of their expanded field of intelligence can expect that the contents drawn forth will involve complete changes in the body. These changes must take place on a transcellular biological level.

To seek the development of consciousness we will require a vehicle that is streamlined, compact, and highly efficient. It must be ready for anything and able to adapt completely to changes in environment, speed, or altitude. We must have a perfect ship that can travel on land, sea or air and be ready to lift off when duty demands. This vehicle is available to us; we are living within its boundaries, but now is the time to cross those boundaries so that new schema of living and learning can develop. We must christen ourselves the unifiers of heart and learn to walk on water with the sages we adore. This lies on the forefront of our destiny.

Magnifying Our Conduct

Our bodies are built on superconductors of energy/ sound/matter transferable to any region, any domain of consciousness. This transferability is unlimited, just as a certain substructure can manifest unqualified specialization. Every part of the body is unique for the job; yet with the proper ticket one can enter into a new domain or habitat without notice.

The study of biology, therefore, must leave behind notions of "primordial soup" in which the organization of cells is chaotic and nonrelational. We enter instead a world which, in matters of structure and function, every cell can be a friend to every other cell. This leads to the possibility of cellular regeneration. Our personal cosmology is the qualifier through which we are made to order; it assures our individuality. The concept of rejection springs from the idea that the human body will not easily tolerate parts of any other living creature. When consciousness can regenerate from within itself there is no concern for rejection of a cell or organ. Our present system of medicine locks away parts for use by others, working with the letter of the law but not the theme of it.

Natural law demands that we hold our own as individuals, creating change through our interchemical matrices of decision. Our own consciousness can grow a new leg, create a new heart, restore a non-functioning kidney. When this is understood, it will not be necessary to borrow used parts from others, nor will it be desirable. When we uncover how to grow our own fields of matter, then we can cultivate ourselves to fit specific conditions, or compensate for trauma or injury. There will be no uncertainty to this, for most of it can be accomplished automatically in an involuntary interface of consciousness and intelligence.

In our present state of transition, we are calling upon certain levels of experimentation to unlock doors of possibility. Some will prove promising, while others, fraught with unnecessary complications, will close shut as we progress to greater understanding. Everything we are doing can be seen as progress, as long as our motivation is out of love; we must not intentionally create harm to ourselves or others.

The Gestation of Living Matter

The control units of living gestation are to be found in the network of pure consciousness as it streams through the interface of our nervous system. The human nervous system holds within its bounds all of the interlays of consciousness, creating spreadsheets of pure energy that form the interlocking frameworks of the body and its subtle counterparts. The body is streamlined through a constant housekeeping process that keeps wastes and unnecessary collections of matter/time from interfering with smooth functioning. When everything is in order, the body functions as an efficient organizer of internal space, rearranging itself in such a way as to promote living gestation. The body makes room for consciousness, thereby providing lighthouses for infinite realization and accomplishment.

Every organ or system is simply a collection of complementary bodies arranged together on a relatively temporary basis in order to fulfill a certain task. Every individual cellular structure is arranged in a precise manner that is unique to that entity. There are no two copies that are exactly alike. This is why self-replication is so crucial to our evolution.

Living gestation provides the means to restore order by creating underlying matrices of pure intelligence that are the read-out sheets for unification to occur at a cellular level. We can create a unified dimension of realization simply by recodifying the message units of the body to perform the necessary tasks. If the body is no longer functioning properly due to age, disease, or lack of proper diet, it can be strengthened simply by relearning the matrices.

This concept presents the knowledge that every individual can learn to read its own makeup. The blueprints of our own bodies appear strange or foreign to us. Until now, our evolution demanded that for the most part everything run on automatic; our intelligence was not highly formulated enough to engage in the process of conscious regeneration. As the curve in consciousness is invoked incrementally, we will become masters of our own craft. This knowledge will be so revolutionary that it will leave us breathless in pursuit of perfection. We will be able to enter into the field of transcendent time in which the ecstasy of pure creation will be available to us.

At present, the act of sexual union produces the fertilized egg which gives us our first glimpse of the power to manifest human life. We see this function as "God-given," which indeed it is, but we are really unaware of how far such understanding can extend. We have created belief constructs that limit the possibilities with reference to creation of life, for fear we will be destroyed. There is good reason for this, since our present circumstances seem to be a direct validation of limitation.

As we enter into a new climate of understanding about the nature of our actual humanity, our ability to create life will be limitless. What distinguishes us from God in this capacity is that perhaps we will never know all of the manifest creation. Yet, managing what is human is our God-given right. Our self-sufficiency, indeed our survival, depends upon it.

Walking on Water

The knowledge that we are capable of feats that are remarkable or even Divine has been with us throughout recorded time. People who eat fire, walk on water, or are able to see or hear over great distances are part of our mythology, yet we think of them as separate rather than as powerful restorative examples of our future. It is time to stretch towards full realization, no matter how different the challenges available to us.

As we evaluate the options for expanded functioning, we will start to develop a language of immediacy for our change into Body Divine. Scientists, especially those interested in organic life forms, are being asked to raise their perceptive voices and proclaim a renaissance in the creative movement of humanity. When we begin to culture the world with the limitless possibilities of our identity, we will unmask the greatness that lies within every one of us. Scientists can offer us this language for change, but it must be remembered that the doctors of the heart are those who embrace life and live it fully, regardless of their chosen speciality. When each person is the captain of his/her own creative voyage, there will not be such adoration of those who have managed to lead the way.

The choice to become a doctor, to explore the options of human perception on the level of the body, is a magnificent one. We must become aware that each of us is capable of restoring order to our individual center in the body. With the proper training, and a new set of tools, the medicine man/woman will only be there to offer suggestions, permitting the person to retain autonomy in decisions of life and death. The transfer of medicinal powers from the doctor to the collective may be the greatest challenge we face at the present moment. Doctors who are bringing the language of medicine into usable and transferable form are to be commended and encouraged.

The Genetic Bubble

Genes are the guiding lights for consciousness to imprint itself onto the field of form. Each genetic sandwich is layered with interlacing pools of perception crystallized into loosely physical form. This form is loose because it is bendable, mutable, and ultimately transferable. The genetic matter that embraces consciousness establishes boundaries so that the infinite can be seen rather than heard. In other words, rather than taking the form of pure vibration, light, or heat, humans have taken the form of matter-radiation. The genetic intertypes we have constructed encourge this chain of being to move along. Every characteristic, every integument that we make, lies within our reach, simply by breathing into the skin of our genetic matrix.

When we can see our genes and remember what formed us, we will be able to make necessary corrections. Faults now occur simply because of stress or change in the environment, which are based on our inability to see the totality. As we view our own wholeness, what we create will also be whole. Any breaks in the union will be subtle and understandable in the new reality. Once we can literally remake ourselves and our children, we will be better able to create value structures which complement our incredible human form.

Our destiny is to create an idyllic center for ourselves and for all the generations to come. Our love for each other will be able to be expressed through compassionate bonding with both changeable and non-changeable aspects of form. We will see all our friends as mates in unified creation. We will treasure all who came before us; they will be honored for making our destiny readable. Every counterpart to ourselves mirrors collective possibility; each will be seen as a genetic brother or sister for possible union. Our concept of male and female will change, allowing for genetic transference of characteristics that we have previously seen as sex-differentiated.

Sexual Strata

What we know as masculine or feminine is a replication of energetic blueprints encoded to represent strata of influence in our development. The masculine component, for example, represents that which binds us to the Earth; it is the thread of human existence that installs us up front on the field of bio-material dynamics. As we become uniform in our ability to walk in all dimensions of time and space, the masculine identity will come into balance with the feminine. Our intuitive, heart-centered nature will be able to align itself with the technological will to create, changing our present evolutionary predicament.

The present social structures are an outgrowth of the hold placed by Universal Intelligence upon our planet until our aggressive tendencies are fully worked out. Once the desire to be here is fully established, we will not be limited by greed and ambition. We have placed structure before function. We have exhibited the archetypal forms and fastened them to our environment, witnessing them as true representations of metaconsciousness.

The lapse in our knowledge is relatively recent. It seems predicated on the advent of technology, but in reality is contradictory to the true feeling that technology could represent. Technology could be a very human face for us, but in mentality we have not progressed much since the initial stages of the industrial revolution. The challenge we presently face is the ability to overcome self-destruction and choose to go beyond it. It may be seen as a game of power in which the bases are built so that we can learn about our tendencies and plead with ourselves to overcome them. Had we taken a different path, perhaps we would not have to play in so deadly a manner.

As we grow into peace, we are confident that we will lose the taste for war; eventually it will become anathema to us. Once this occurs, we will have little need for such a clear distinction between the sexes. The unisexual train of being rides upon a wave that is more promotional in nature. Our separateness concerning sex roles and even the nature of gender have been based upon our need to accumulate territory and to acquire food. Since not everyone could engage in hunting, some part of us had to be reserved for tending the fires, raising children and staying home—a role which was for the most part, traditionally given to women.

Breathing the Masculine and Feminine

The masculine and feminine constructs are the building blocks for our present understanding of complementary differences in human life. When we breathe the feminine we uptake the receptive delicacy of feeling that is responsible for our ability to live life in a natural way. The feminine force has been represented by a closeness to nature that seeks to unify with the feeling of the environment. When we embrace the feminine, whether housed in male or female bodies, we begin to encounter the depth of our receptivity to need, change, and purpose. We become listeners in the interior dwelling place of collective consciousness. We learn to mother ourselves and nurture those seeds that are the most creative and promising.

The masculine imprint represents the glue that binds us to the formation of matter; without it we might be housed in spirit rather than in physical form. The irony is that although women hold the honored stature of bearing children, such creation could not be possible without male linkage to the formation matrices that bring matter into recognizable vision. All of the stages of matter are brought together through male and female union, and can be pulled apart only by holding to their separate interfaces. When female and male unite they are simply imitating a union that is already established within them. At the present time, sexual differentiation allows us to comply more fully with that which we truly are.

Why We Dance

Every construct that shapes us can also break us. This is the Universal Law. Out of the threads that pull the fabric of the body together are cross-weaves of both male and female consciousness. Even a child born without a physical father, which might now be biologically possible through what we know as a virgin birth, has an internal male value.

For the most part we view the difference in the sexes as a combination of heredity and social culture. It is difficult to determine what is truly masculine or feminine. Perhaps we can understand the situation more easily if we think of it in terms of shaping and unshaping. Every matrix of consciousness is fed through a field of unstableness that is fluid, non-formed, and inconjugant. This field, which is relatively static in most of us at the present time, binds us to the force of structure, and allows us to appear as we are.

We might say that in our scheme of things this unshaped value is more feminine. It is looser, more refined, and more easily changeable. The masculine is by nature more immutable, because it buckles us in to the seat of consciousness. The mutable, feminine form is attracted to the more immutable masculine form, whether it is in someone of the opposite or the same sex. This is because we are seeking balance, and balance happens only when the formless and the full of form hold together.

As we become less dependent on having to hold ourselves in place through fixed constructs of understanding, the need to have such complete differentiation and thereby so keen a structure of the sexes breaks down. We see this now in our social functions, but these unisexual tendencies will soon become physical as well. As women become more structured, firmer and more capable of physical strength once reserved for men, men will be able to become softer, less controlled and more capable of seeing within the matrices of their own consciousness.

These changes may result in two sexes or simply a great range of sex-differentiated beings, constructed to fulfill attributes of structure and function in alignment with purpose. Perhaps no one knows the final outcome. It seems clear that the condition of fixity in our present sexual identity will continue to be broken down; the social forms that reinforce it will also undergo tremendous and rapid change.

Holding Ourselves Together

Every body has a memory, and every memory is based on the backwards and forwards uptake of information as it is recreated in the field of the body. The body laughs at itself, looking into the mirror of its continuation; it is continually fed messages that allow it to bond to matter and to repair itself. The degree of ability to glue ourselves together, the type of bonding influence that is created, makes us what we are.

We are the product of magnetic forces that attract and repel in order to form matter. Every body is composed of fully enhanced crystalline matrices that are both uniform and nonuniform. They complement one another, providing adjacent threads of being that bind us to other lifeforms in our world and beyond. We are color-encoded organisms that are bound to all other lifeforms by the chain of command of pure consciousness as it registers in our DNA.

When DNA speaks to us, we reveal our presence by letting it be known that we are of a certain shape, size or color. All of what we are is lifted into matter through the DNA mouthpiece, but the degree to which it accurately portrays what we could be is rather small at the present time. Our DNA has become the *wraparound value* for pure consciousness. It allows us to matrix-point ourselves into the level of physical stature.

As our DNA loses its charm with our present array of performance, it will reconstitute itself to present new possibilities of development. The present statutory changes in the laws of matter that are going on all around us are providing opportunities for our DNA to redirect its functioning to more evolutionary spirals of consciousness. As this occurs, we will be able to dance around the tree of life in such a way as to unite all of our bonds with infinite perfection.

DNA: The Medium of Assembly

The DNA avenue of repair lies in its wondrous ability to assemble itself in many different ways with the same essential coat. Unlike other species in our sphere, we are infinitely able to be reassembled and repackaged. We can come in many colors, many tribes, speak many dialects, have many different types of nervous systems, and be essentially and fully human.

This offers us a great advantage because we can speak the tongue of our brothers and sisters without having to develop a different medium of existence to do it. We can call each other up on the communications line of consciousness, and we can document our understanding of what we share in very great detail. Our perfectly transferable and reliable DNA housing allows us to do this.

DNA provides an interchangeable language of admissibility to us. We are able to return the coin of consciousness to the phone. When we stir up consciousness and cause it to provide us with its primary imprinting, we are letting DNA arise in and of itself. It becomes the bonding mechanism for consciousness to return to matter and vice versa. DNA signals us to awaken to our individuality, even as it allows us to have comprehensive communication with other lifeforms, particularly human.

When DNA is speeded up, as it is now, it starts to undergo a new itinerary of bonding formation within its infinite spread. DNA becomes capable of lifting matter into new vibrational patterns of integrity. We are literally lifted up, emblazoned with the patterning of our ancestors in addition to being infused with the unknowable. The Host Intelligence makes room for creation in our DNA superstructure.

The retention is accomplished through chromosomal activity that allows matter to be tied to its stream of intelligence, making human beings capable of revisiting their ancestry. We are installed on the seat of leadership, even if this leadership is limited. Once leadership is met by pure reason, we are not limited to what our DNA has conjured for us through the happenstance of causality. We are able to make and break forms, thereby reconstructing the genetic blueprint of consciousness.

Every synchronized chain of amino acids will be able to step into time by empowering consciousness to revolutionize itself. Matter will adhere to the causality of genetic representation. It will become capable of creating purely organic forms that function much as the machines we know now. Matter will become more precise, more complete, and more capable of imitating consciousness in its fluid and creative state.

When we are able to create ourselves out of inner matter and extend this range of creation to our environment, we will enter an age in which inner and outer will become interchangeable. Until such time as this occurs, we can only move along on the field of DNA enchantment, gasping at the forms we have created and mimicking their possible outcomes.

As we play with the bonding mechanisms, we reunite ourselves with the full field of parameters; in so doing, we are implementing a steady curve into the field of mutation. Rather than relying on a change in species, we have come to rely on a change in continents—the actual infusion of matter that is earthbound into our physical structure. We are literally becoming the mountains, streams, and air. Our physical earthform environment, which was and always has been representative of the body, will become our bodies. There will be literally no distinction.

Since Mother Earth is our actual causal body in consciousness, we will give Her more power in engineering our primary design. When we attune ourselves to Her advantages and understand Her prerequisites, we will strip ourselves of the unnecessary and will move into a consummate union with Her. In this sense, we will become Earth children, even as our goals become more interplanetary and biorelational.

Photokinesthetic Functioning

In present physical terms, we are bearers for the infinite valuation of light in our galaxy. As we propel light into our energy formation, we encircle it with matter and spin our own body identity from it. This use of light is the primary mechanism by which we are able to qualify consciousness and allow it to become part of the body.

Our present system for the uptake of light is limited. We breathe light so that it is managed through our energy system, rather than through a direct light/energy/heat transformation. Plants in our sphere are far more direct in this regard. Their ability to create life through the process of photosynthesis allows them to take in light more directly and to uptake it more efficiently for the production of matter.

Our experience of life is thought-generated. For example, what we think about our climate affects our ability to adapt to it. The plant functions in an entirely involuntary way, and its uptake of information is photochemical, rather than thought-responsive. We think of our breathing as completely involuntary, yet we are perfectly capable of stopping it. The fact that we are able to take our own lives is what makes us special, but it also is representative of the degree of conscious control limiting our possible involuntary expansion.

In order for us to change our relationship to light, we must think of it in different terms. We now know that we crave sunlight in winter, for example, or enjoy a fine day at the seashore, but we are not completely conscious of how essential our uptake of light is for proper formation and functioning. The light spoken of here is not only the physical value. It is the nonphysical or cosmic value of light, the generative force that illuminates the universe. The harnessing of this force, and its applications to our environment, is essential to our present development. We need to experience light as a force-field of immense proportions, crystallizing through physical matrices and thereby entering our environment. In this way, we can start to perceive a boundless light that is manifested on both inner and outer levels.

As the quality of interior light expands in our perception, we are able to think less monovisually and more radially. We become imbued with light and begin to see it as a radiating energy that permeates our cells. In this state, we can draw on light to develop new styles of energetic functioning on the physical level. We can restore the body by enveloping it in crystalline light matrices which hold the pattern for recreating centers that are in disrepair. They can then be fully reinforced, becoming capable of self-healing or self-replication.

In addition, our ability to use physical light, to ingest it and develop it economically for our own benefit, will be a great advance in the near future. This includes the common use of solar energy for travel, heat, etc., but also our use of it for internal fuel in our own makeup. High chlorophyll foods and the use of certain organic chemicals that boost the photoreceptivity of our collective consciousness, can become available to us. Through this means, our entire environment will become advantageously solar-related.

When we mix the fuels of our labor on both inner and outer levels, we will bridge the petrochemical barrier and not have to rely so much on depleting the body of the Earth for our needs.

The stability of our present planetary structure depends on our willingness to unify with the light and to be able to render it for our purposes. This must be seen as literal rather than figurative.

Lighting the House of the Body

When the body becomes illuminated through an intense bonding to the light, it is capable of scanning its own functioning; it can remember the switches that turn on particular energy corridors. Before this occurs, it is as if we are living in an unlit house—we cannot see our way around. We bump into furniture, duplicate unnecessary actions, and are unable to see how to put things together. Once the light is on, we can become more focused; we know what jobs there are to do and can maintain a working system of light that manages our physical situation.

As we start to operate more consciously to develop interior mechanisms for synthesizing the body, we enter into an organic union expressing who we truly are. We can view ourselves as an interlocking matrical system of moveable parts, each with a specialized function, but each capable of learning from the other. This changes our relationship to our bodies very dramatically. We can then enter into concert with what was once involuntary.

The primary example of this is our ability to work with the physical breath to change our personal course of action. We can program the body to release certain toxins or foreign material and can talk to ourselves to produce new, more advantageous directions for consciousness to travel. Once we can link with pure consciousness, looking into the road system through which it can travel, we then must learn the grid of our own awareness. We re-educate ourselves about the framework of response that generates a whole being.

This relationship to our wholeness occurs on both physical and non-physical levels, creating a marriage of form and spirit which is the fundamental arc that creates the possibility of shape-shifting. When the consciousness can turn the body around and rework it in the keenest detail, we have gained a level of mastery over matter that is completely unique. Our cues become clearer;

the individual hold on form is loosened, and we can reprocess ourselves to adapt to changing needs.

At this stage, we can link directly to the time matrices that feed the influx of consciousness to the body. Once this is established, we can live in a complementary value of time that is constantly being recalibrated to fit our present moment. The body/time field moves with us across the active relations of our life and becomes interactive in every domain. We are alive to ourselves and are lively to our environment. The period of static relationship comes to an end.

Cellular Feedback

Consciousness frees itself from matter in a way that is unique to its own expression. Consciousness, by wrapping around the box of creation like a cloak, has the opportunity to peek inside and determine its contents. The physical body is composed of germinal matrices that we call cells. Each cell has its own specialization and is complete unto itself. When a cell becomes incapable of perfect replication due to age or disease, we try to destroy it, hoping it will not spread its negative influence to other more perfect counterparts. We have very little understanding of how to create a new field of development for the cell to find its way back to its origins, thus we believe our only solution is to eliminate it. However, we never really eliminate the cause behind the inappropriate display of life that the cell represents.

Cells communicate with each other. Every cell has a photochemical coding mechanism which creates light-sensitive heat responses that directly transmit information from one cell to the other. Every cell therefore is completely aware of what is going on around it. Cells, however, do not react the way we do. They do not interpret their signals as words or conceptual structures, but instead telepathically address their needs to each other. They provide photo-imprints of what must be done and then wait patiently until these imprints are accomplished. If this were not true, our bodies would disintegrate because they would have no inherent memory of their construction. In a body without photomemory, the lessons of reconstruction would be unavailable and we would die out within minutes of our creation. Instead, every set of cells has its own instructions for replication and goes about the business of restoring order in every moment.

The communications system used by the cells is now completely nonconscious for most of us. We are not linked into the memory network and cannot see the communications taking place. Once we become more aware, more infused with the light value, we begin to be able to actually see ourselves in operation. As this occurs, the cells link more directly with consciousness and in a sense become conscious themselves. They link directly to the central nervous system and are managed by our minds. We are then able to create brain-cell links that can allow us to recalibrate our cellular functioning. Once we can communicate directly with our own personal predication system, we enter a world in which timing, and the understanding of life-cycles in great detail, is available to us.

Numbering the Regions

Life is a mathematical function. Mathematics, however, is not a function that lives outside our interior responses. When we begin to live in cellular awareness, the system of mathematical responses that allows photochemical communication to occur becomes available to us as well.

Life is a process of ratio-synchronization in which all of the values of consciousness are in relation to each other; these relations are qualified by mathematical variables. They can be studied, enhanced or replicated. Once we understand the vibro-chemical relationships of matter and function, we can develop inner communication that allows us to see mathematical corollaries.

This type of mathematical superstructure is a felt/seen response. It cannot exist purely on the level of thought or reason as we know it. When we can begin to think mathematically by actually viewing the constructs of identity, we can then alter the ratio-synchronization of movement on the level of the body. The body talks to us in mathematical terms, and we can respond to it.

The realization of the numbers is developed through seeing the cross-matrices of time as they express themselves in the field of the body. We can actually view time as it stretches itself over the field of matter that is our own body. This appears to us in mathematical equations or calibrations, which can take on symbolic terms, but are in fact purely literal and nonrelational in their essence.

Since time is both static and ever-changing in character, this type of mathematics expresses the absolute or causal value in relationship to the changing value. It is a mathematics of incremental restitution on the level of form. The ratios involved are the calibrations of changng and nonchanging movement as expressed through field/time generation.

We ingest this mathematical infrastructure, interpreting our actions so as to understand the reasoning behind our entire selection of mathematical functions. We then become self-selecting in reference to our ability to exchange forms.

The beauty of all of this, in terms of our evolutionary development, is that it happens completely spontaneously. We cannot order ourselves a mathematical dinner and ask it to be served at our table. The consciousness itself breathes life into its surroundings; as it does so, all sorts of curious data appear. We can only observe this phenomenon, and as it transduces, learn how to manage it. We are witness to the development, and as we grow accustomed to it, we can interplay with it. We are involved in its power, which is essentially beyond us at its maximum output.

As we determine our causality, we can also see more clearly what we will never determine. Becoming more conscious, we are willing to trust more of ourselves to automatic response. By this time, it is out of choice rather than out of ignorance that we turn ourselves back to the involuntary. This allows us the capability to look in at any time, when it becomes necessary. We can wake up to the Divine order without feeling subservient to it. It is therefore a matter of psychological change as well as physical.

Assessing the Status of the Body

The health of the body is determined by creating a balance between the Host medium and the physical manifestation of intelligence. The organs and systems are points of reception for the interchange of intelligence and are maintained through the impulse generation of the healthy nervous system. The body is the housing through which intelligence can express itself, decreasing the rate and formation of input so that it can be interpreted to produce fundamental activity.

The health of the body cannot be measured solely in terms of its physical flow, but must be assessed through understanding the wave-point junctions of all sectors, once they are solidified on a psychophysical level. To establish boundaries for assessing the physical integrity of the body, we must determine wavelength attributes in the body with relationship to three factors: 1) the status of *impedance variables* responsible for wavelength output; 2) the rerouting of Host information so that it is in correspondence to the repair mandibles for the flow of information to the organs; and 3) the increase in the space-wave function diameters in connection to the actual free-space that the nervous system provides for signal entry.

The body functions as a wave-transformer information system for intelligence. Codes of light/heat/wave transfer are taken up through the systems and fed into the central control of the human brain and nervous system. This intelligence is then tracked for output to the organs. Each section of intelligence is upgraded with respect to the quality of intermediary bridging mechanisms the organ is capable of generating so that it can be switched on to the primary feed from the Host Intelligence.

Without this function of wave-generated impedance, the signal would actually loop around over the organs and not land properly; it might not be seated in such a way that it could be taken up correctly and fed into the cells for proper correlation. The necessity of downgrading the signal is what constitutes the principal problem with relationship to functions of physical health, since they are essentially translations of the primary intelligence rather than a direct and nonimpeded impulse variable.

The organs are stripped down to their essential chaining mechanisms in the ribonucleic center and not allowed to function as consistently or compactly as they might. As this occurs, they lose their ability to individualize within the context of their own process of differentiation. Allied with the field, but not spontaneously attuned to its entire chorus of signal variables, there is a tendency to rely on secondary rather than primary methods of organization, thus robbing the cells of their primary means of "cultural" integrity.

In this context, the cells are weakened; they are no longer capable either of functioning entirely on their own as principal aggregate matrices nor as combinations of cells differentiated for specific purposes within the functioning of the organism. This situation leads to rapid confusion or weakness, preventing healthy mitosis and robbing the cells of the intelligence needed for smooth functioning. These conditions can often give way to dis-ease manifestations such as cancer, which in a sense is a form of arrhythmia; the cells no longer have the proper timing mechanisms to loop-lock over their own interface and develop perfectly synchronized "buddies" in their own field array. Stranded, looking for others to bond with, they link up to cells that are not properly prepared or seated in the synchronous blend of differentiated homogeneity necessary for proper matching.

This weakens the cells in their aggregate formation and causes them to bunch up in tumor-like formations. Cell repair becomes difficult, since the improper bonding causes the cells to forget their original strings in the cellular grouping, losing their mechanism for repair. The intelligence stream that creates the proper signal formation is often irretrievably lost. The only remedy for this would be a type of false bonding in which cells are introduced that are healthfully following the new plan, relaying the messages necessary to host cells for recongregation. This must occur through naturally integrative means rather than introduction of tissue or organs from foreign areas of the body.

The body that enters the vaults of time and up-puts its signaling mechanisms so that more untouched aspects of intelligence can enter, will remain healthy and vigorous for longer periods of time. The healthy body is constantly changing its potential feed in response to the intelligence signals and can stay on line all of the time. Even during the rest periods, the healthy body is directly online with respect to the signals of pure intelligence and is transducing them only so far as it is necessary for them to make sense in a physical way. The less this signal output is touched, the less static the body is, and the more capable it is of remaining fluid and loose. Once consciousness has been stabilized at higher vibrational outputs, the susceptibility to breakup lessens, and the organs are able to hold their positions more efficiently with respect to consciousness.

The Globobiotic System

A curious phenomenon emerges when many human beings are able to congregate together who are in the process of upgrading the signaling mechanisms as they are fed into the centrifugal flow of the body. Each of the bodies becomes its own relay station for a particular channel of information, much the same way as each system of organs would ordinarily do. In this sense, the group becomes one globobiotic system, with certain individuals holding the function of respiration, digestion, reproduction, etc. Once the signals are housed in more than one conduit, they are able to further strengthen and differentiate themselves. Then, at a certain point, these signals can be relocated or refocused into a single body, becoming more capable of taking form once the successful relocation has been mastered.

This is the reason why experiments in group collectivity are so important. A free-moving, advanced set of parameters is established, opening the doors to the output of higher intelligence. This can then provide an essential grounding of ideas which might not ordinarily be possible. The upgrading of intelligence through the response medium of a group mind is central to the transitional phase of development, at least until such time as our nervous systems are able to coast along on their own, with respect to signal uptake.

As we expand our awareness, the group can then function as an intermediary form between the collective godheads of the Infinite and ourselves. We can thereby locate ourselves with respect to other arrays or fields of globobiotic consciousness, providing barriers in ourselves for certain areas for which we are not prepared. Such barriers are satisfactory as long as we have others who can take up the slack in a given area and feed us the status of information.

Once we understand that intelligence can be collectively generated, we begin to develop degrees of specialization without losing sight of the wholeness of the primary act, namely the uptake of the "pure consciousness windows" through which the time/healing matrices can establish themselves. The possibility of limitless union therefore is expanded to include a "cell" of individual entities, functioning similarly to the individual cell and with similar needs and properties.

Interactive Fields of Intelligence

The body crisscrosses its own blueprint, establishing a center of operations with others with whom it is compatible. This develops a flow of information that allows for refined assessment. The individuals can then fuse together to identify fields of movement within the environment. They can act as conductors for the flow of information from the body of the Earth to the collective mind.

There is an unfounded fear of losing individual identity, since the collective can function well only when the individuals are highly educated, easily adaptable, yet infinitely specific with respect to their needs and desires. The range of motion of the individual is enhanced through this collective action, and the possibilities for fusing with the mind/field of the Earth are enhanced. These unions open us to true perception and decrease the need for inhibitors.

The possibility of direct communication with plants is greatly increased, as well as listening directly to the wave formations of animal species. This provides intercedent waves of communication for the purpose of evolution. The possibility of communicating with all species of animals has been hindered by our unwillingness to grow together for common purposes. Once the fear of this activity is suspended, enormous freedom will result.

Cellular Housekeeping

The Body presently houses much information that could be discarded to make room for uptake that would be more conducive to growth. To house the full range of regenerational information, old memories of past faults or limited situations should be dissolved. To do this, the individual must sort through his/her personal history on a memory uptake basis through actual physiological processing. This cannot be done simply on a psychological level, although such "looking at the books" can enhance the physiological process.

The body must be cleansed of stored memories and this process is best accomplished through the use of natural substances, particularly from the herb and animal kingdoms. This is because those species available to us in the plant and animal worlds hold key information about the processing units of intelligence; these can be swallowed whole and interpreted by our own central nervous systems. These information units are already streamlined for living use and are readily available to the cells for understanding. The synthetic varieties of such substances can be helpful; however, they do not have available this living program of knowledge and they must rely directly on the body's own accumulation of knowledge as to how to utilize them.

A body that is in disrepair is limited in its scope and function, and is also limited as to its operating instructions. It therefore cannot derive a holistic perspective for the synthetic substance as far as usage and propriety are concerned. We might say, then, that it is not the synthetic substance that is at fault here, but rather the body's inability to know what to do with it.

The herbal or natural substance, on the other hand, contains within itself a set of instructions. When it is taken in on a cellular level, it provides the proper series of breakdown functions that allow intelligence to be uptaken directly into the cellular Host parameters of the body. Once the body has scanned the herbal substance, it is capable of delivering it to the proper area. Even if the substance is ingested and/or applied to one area of the body, it will quickly be relocated to the area that most needs it. In addition, the natural substance from plant or animal derivatives acts as a carrier of information in a way that permits the imprint of intelligence to last for a longer period of time.

Once the organ has seen itself in the shape of the plant or animal matrix, it literally takes a picture of it, and this picture is implanted in the memory of the individual human cell. With this memory in place, the individual may no longer need to take the substance physically; the coding mechanism for its usage, and the actual information with regard to its proper manufacturing has been delivered to the cells. They have been given the gift of operations and will thereby be able to use it.

The cells are fed from these natural models, beginning to function as the plants or animals would do themselves. They become involuntary Host carriers for the signal until the brain and central nervous system can make sense out of the new information. In the advanced individual, every substance can literally be translated into a type of thought apparatus that promotes replication, regeneration and centering of intelligence. This becomes a very enjoyable process and can lead directly to full immortality. Once we no longer need to ingest natural substances, we can essentially grow our own. We may simply need occasional reminders.

Weighing the Elements

The primary element of the human body is that of space, or the interior return mechanism of consciousness moving to its primordial sound value. The interstices of the body/space dimension hold the underpinnings of free-floating consciousness as it travels or swings from solid to fluid. The body is a spinning wheel in which threads of consciousness are created, woven together and then mapped in order that breathing room can be established.

Every body has a different conception of its interstitial makeup. It cannot change the basic drawing plans of spatial arrangement until the solid quality can be fully charged to accept new registration as a nonsolid mass. In a sense, the body would be weightless were it not for gravity holding us to the Earth. The body itself is not really heavy in its own right other than its naturally fluid or loose state in physical matter.

When the body starts to transmute itself and develops its full shape-shifting potential, the entire nucleus of each cell undergoes a tug through which it is allowed to stand freely in its own base, thereby creating bonds for the waiting chambers of consciousness. It is as if the nucleus stands at attention, waiting to be attracted by the forces that can hold it together. The centripetal forces that spin the body at the center help it to maintain its fluidity, even as the pressure within the cells themelves becomes more self-contained and rapid. As the "pressure pins" are pulled, the body becomes more lightweight, buoyant, and able to carry its own celestial requirements at great distance. It can speed along the consciousness highway and maintain its own dignity and poise without breaking up.

The centrifugal forces that pull the body apart and appear to push it to finer levels of compressed matter, act in conjunction with the centripetal spin of the body's inherent granular structure. The particle/wave formation of energy enters the physical body and creates an interweaving of centrifugal and centripetal movement that essentially compresses the impact field without dramatically loosening its essential character. This lessens the facet-weaving of interdimensional change so that the psychological character of the remaining personality can tolerate the rapid advancement towards unity.

The body is then bound together in a fine-weave or consciousness mesh that keeps it lively and continuously forming its interstitial map. Having withstood the possibility of breakup, the body, as created through the ingestion of new quantities of super-data into the matrical fields, goes through a cross-breeding that causes it to be able to change species. This species shift is in the direction of a fuller flowering of humanity, rather than a break from it.

The new human being will be able to create a magnified vision of perceptive consciousness so that the body, while still remaining weightless, will have a more definite form. This seems like a paradox, since the free-floating, shape-shifting body described would appear to be less solid, more fluid, nonstatic. Indeed this is true, but the quality of actual weight, actual concretized matter, could in fact be greater; more of the true consciousness of the human entity can be developed into physical form once the perma-nature of the body has been shifted. Consciousness is allowed to stretch itself over the full weight of the body and will in fact create more definition for ourselves. We align ourselves more solidly with the universal matrices that hold matter together, thus becoming more solid though appearing to be more liquid. This will create a great difference in our powers of material manifestation.

The Way to Creation

The physical body that is made more solid through this process will become adept at holding to that which is of itself, i.e., all of the matter which is connected directly to its nature. Since the transformed body will paint the pictures of the Earth through its own prismatic resonance, it will operate fully and completely in unison with the Earth. The pulse of the Mother itself will be its own pulse, with little differentiation on that level. The differentiation will come through perfect alignment with the Earth and all of the cosmic forces which bind the Earth to this section of time and space.

Once we have bound ourselves back to the Mother, we will essentially be more free, because we will become part of the global body. This is globobiotic consciousness. Once we are part of the skin of the Mother, we enter into a free-fall continuum with Her, making it possible to travel through time and space on

Her wings. The growth of collective consciousness in this direction will allow us the benefit of enjoying spaceship Earth, a free-swinging jag into the heart of the space-time continuum.

When the Earth itself is no longer locked solidly into a static dimension of time for the sake of making us comfortable, it will be able to be interplanetary in its movement. This may seem startling, as one imagines that all planets are fixed in their dimension by their own bonding mechanisms. We imagine that the laws of the Universe insist on planets that spin happily around larger bodies until destroyed through accident or lack of direct sunpower. The reason we have this image of the Universe is that we assume that planetary behavior is a holding pattern in time and space and that civilizations meet each other through a rigid encounter with these universal phenomena. Once a globobiotic consciousness is truly in place, we will not have a permanent home, but will be able to transmit ourselves anywhere we wish within certain parameters; even these paramters can stretch as energy and consciousness demand.

Therefore, it may be accurate to say that the Earth itself will not have to remain fixed in her blanket forever but will instead become a companion traveler. We will take a slice of the Earth's fundamental matrices with us, because in an absolute sense, we are the Earth. We will be voyagers carrying our cosmic identity throughout the Universe.

As the Earth shifts in its swing, we will be able to create a different understanding of life, destiny, and our means of development. In this sense, we may think of planets as going in and out of phase interdimensionally in a similar fashion as interstellar spacecraft do now. However, we may find that our ability to move through time will preclude the need of physical vehicles, except in cases where the interstellar units need to travel very vast distances. We will be able to create waves in the field through our own consciousness and be able to transmit ourselves to any point on the whole plane. This will involve sharing the Earth with others and will lift this dimension into the field of a whole value of interstellar light presence.

As the body exhibits more uniform dimensionality, it will present itself on the field of Divine forms and will caress the Infinite. As this occurs, its transformation will not be just a singular event but will be never-ending, vast in its consequences and perspectives. We are at the beginning of this body/mind gate into other lifeforms and their place in the Plan.

The Tent of the Body Infinite

The body that is motionless, yet perfectly capable of infinite motion, that is breathless, but perfectly capable of swallowing the planetary rhythms, enters a state in which it can create a tent or blanket of perfection. We are then capable of integrating the matrical influences of consciousness and can skate into the "ice-planes" of space-time awareness. Balancing on the foot of the Infinite, the body will be able to swing itself to the left and right to establish boundaries; however its poles will remain loose and fluid, much the same as a tent. It will have places at which it is held, but these will be movable structures, capable of being carried off and reset at any point. The body that can conceive of distance as a function of motion and motion as a function of time, will enter the free-fall continuum in which everything that presents itself before it becomes part of the play within the field. All is manifested through the ins and outs of form via the cognition of availability.

The seeds of the Body Infinite lie in the present DNA matrices. Each shape of DNA is filled with mouthfuls of crystal continuum which, when swallowed through the process of cognitive suggestion, strike the magnetic resonance of human form and are made manifest. They will not be limited by generations of DNA formation that we have seen previously.

It must be understood that the individual entity has breathed identity into itself through many, many existences. These existences are not fully understood by us. Some of us conceive of these forms of expression as lifetimes, which they are in a sense, but they may be lifetimes of conjugant expression which may be symbolic rather than literal. The casting into form comes only when it is absolutely necessary; just as we dream thousands of dreams a week, the dreams of existence happen much more swiftly and fully than the time we have to manifest them. We imagine ourselves through our lessons, our discouragements, our triumphs, much more often than we literally experience them.

If this is the case, what is life itself? Life may be viewed as the primary crystallization of matter/time into the space-time vault, the stirring up of the interstices so that the gaps can be made more uniform, more relative. Once these gaps are organized, restructured, reprinted on the map of awareness, they cycle back into the primordial whole only to be shaken again and reconstituted in a different form. It is much like the combinations of dice on the playing field, with very complex mathematical consequences in the structuring of the physical and biochemical formations. As these combinations are introduced, the light forms can study themselves; this is what we know as consciousness.

The study of the worlds goes on at all times and places in the dimensional bridgework, even as our seemingly limited existence is brought forward into the life plane. We are layered like sandwiches of pure dimensional consciousness, brought together for particular purposes, fanned out in all directions and returned to the inner eye at an expressed interval. All of this goes on in the bakeshop of the body. This is where the temperatures are just right, the mechanics in good working order. We are cooked, made available for certain forms of inspection; we become the sweets of the Infinite. We are the tender cakes that consciousness bakes, encouraged to rise even when we may feel that our life experience is lacking the momentum to do so. The forms of expression we take, no matter how meaningless, traumatic or dense in manifestation, are brought together on the level of the physical body.

Sacred Bonding with the Stars

As the body develops wings for travel on an interdimensional level, it is unified with the sacred light that is its birthright. It enters into the parallel space-time vantage point from which it can look out at the stars and expand itself to the outermost regions of its own longing.

This infinitely expandable body is capable of interdependency with the Host Intelligence in rather unique ways. As humans, we are currently subject to all the ills that human beings are heir to, namely, the aches, pains and other more acute discomforts that arise as the body ages or falls into disrepair. As the body is released into the Host, it develops a means of infinite expansion that allows it to wrest itself free from any impairment that is not completely in alignment with its wants or desires. From this mantle of valuation, the body becomes infinitely restructurable.

The beauty of this process is that the possibility of restructuring exists at all *frequency response levels*. In other words, as the body shifts into high gear in response to the manifold possibilities of creation, it can replicate itself at will, as well as duplicate any possibilities it sees that were not part of its original expression. In this sense, the body becomes the efficiency leader of the Universe, seeking refinements for itself from the matrices of consciousness it reaps from the environment.

As we imitate our friends in the most economical way, we gain a sense of limitless proportion with respect to variation. This cross-generational capability affords us the possibility of marrying among the star races; i.e., the ability to germinate our population from other dimensions of intelligent life. Once we rearrange our definition of what it means to be human, we can be friends with our star neighbors and not hold to our own bonding so closely. We can intermingle successfully with all of the Host Creation and not fear that we will be despoiled by our commonality. It is a time of sacred bonding with the basis of life itself.

Our training in the meantime, as we seek out the reaches of our union, lies with our ability to maintain variation and understand similarity. As the body shifts to an identity that is consciousness-generated rather than a repetition of previous responses, we can live every moment refining our lives for the possibility of variance. In this variance lies our future.

Adjusting to the Now

As medicine in our sphere becomes directed less towards the cure of disease and more towards the advancement of intelligent life, the shift to a refined sensibility in its practice becomes evident. Every physician will know the cure, and this cure lies within the grasp of every individual. It is the cure of originality, the ability to manifest possibility through cross-dimensional intersection.

When a physician becomes a model for redirecting his/her own consciousness, the patient is inspired to follow. This is why physicians hold an important place in our present timetable. The sacredness of the body as a passageway into consciousness must be more fully understood, and the physician has the value point to instill such knowledge by unmasking the true nature of living organisms. When the nature of life is known in its nonchanging and changing values, it can be revealed for what it is: a miracle that is brought forward through the memory of every individual participant.

The physician can develop procedures in the direction of unification, even before a whole value is completely within his/her grasp. To do this requires bravery enough to set aside previously stored understanding about the nature of the body, and develop a wedded seal with his/her own inner life. In so doing, the technical mind becomes a bridge to the unified mind. The scientific framework has always left room for the unknown; it is simply a matter of opening the mind to its own nature.

Medical procedures that emphasize naturally occurring substances, such as whole plant foods or other organic compounds, are more in keeping with the concept of living memory. The value of cutting into the body through surgery is limited, but this procedure may need to be maintained as long as the physician is incapable of seeing how the body can cut into itself. The body is a very aggressive place in terms of temporal activity. It goes after invaders, strips away what is unnecessary, and is merciless in its deterioration if its needs for consciousness regeneration are not met.

The physician need not see advanced expression of the body as a process that will lead away from his/her calling. Instead, the physician can express a feeling of extended purpose, a perspective on human life from an exciting and intangible view. The physician will in the future come to rely on a team of intelligent counterparts who will assist in research, methodology, and development. As the body is more clearly understood, and its processes from the level of primary genetics to total restoration known, a very advanced system of cataloging with respect to medical records will be required. At the present time, we have more records of the condition of the vehicle we drive than we do about our own bodies. We utilize medical technology when we are sick, but know little on an everyday basis about the entire tissue structure or function. Sometimes only death provides information on this level.

Those who study the structure and function of the body might be invited to research every aspect: begin with the Self, understand the individual chemical and interlocutory makeup of organic function and describe what is seen to those who will listen. The new technologies will have to be generated through each individual discovering what he or she is made of and then attempting to explain this knowledge through technical means. It is a complex system of Self-research and will require the collective mind to accomplish it.

Doctors in studying disease often study the field of death rather than the field of life. It is similar to studying the weeds in the garden instead of the flowers. However, the knowledge of the weeds and how to eliminate them can be readily translated to study the principal members of the garden. The transformation is from a construct of fear, preventing the onslaught of foreign matter, to a construct of researching consistently novel approaches to creative imagination.

The financial barriers which limit medical research are placed in our consciousness simply through habit. As our priorities are redirected towards longevity and evolution, the physicians, who are traditionally pledged to preserve life, will be on the forefront of creating it. All those who work in allied fields can become helpful members in this drive towards perfection.

Circuit Breakers in the Electromagnetic System

Since everyone can be considered an organ in the collective matrix, one is also one's own switch which can turn off certain sensibilities even as they are created. For this reason, the person who has redefined his/her body increases internal fluidity. The individual opens into certain avenues of consciousness which function as light-drivers in the collective turnaround. When specific frequencies are circulated through the physical body for this purpose, the individual loses a degree of response-generated identity, becoming a conduit for advancement in his/her sphere. The work of the switches is akin to the ability to incur grace through release of the personal "I," breaking the field of motion in consciousness and establishing the body on another plateau.

The body that is functioning for and through collective purpose elects its choice of medium. It is friendly to itself through its own parameters but is also friendly to those with other kinds of parameters. In this sense, it is complementary to its environment and all of the individual units around it. The body that can restore order to itself and dedicate itself to the restructuring of consciousness on a more than personal basis becomes a different type of member in the matrix. It becomes a switch for higher frequency energies to be charged up or down through its own space.

As it enters into this collective harmony, the system of checks and balances keeping the individual organism responsive to change may fail. This is because the complementary body must function from the viewpoint of a team consciousness. It becomes dissatisfied with function or action that is not directed entirely to a more than personal goal.

It is important to understand that this change is biochemical as well as psychological. When the mind functions from a transpersonal perspective, it creates a different framework for the functioning of the body. Unification must constantly take place by the body with other generational beings who are on the same or complementary response frequency. If this does not occur, illness may result, because the individual cannot maintain the full range of frequency response generation without hooking up with his or her counterparts.

The system prevents overload by insuring that each individual who goes online with the evolutionary matrix is ready to create a level of psychological and spiritual dedication to the full range of causality. Otherwise, the split in identity between the personal "I" and the generational "we" would be so acute as to literally destroy the functioning of the body. For this reason, the medicine of the future will be more fully dedicated to studying the interdynamics of life rather than individual responses. Self-differentiation will not normally be accomplished simply by the Self, except in cases when such research or understanding is imperative to the whole. As medicine shifts into this type of formation, order is restored to its interparticipants—circuit breakers who, by their own inner development, will be capable of taking on more of the load.

This might be seen as similar to those advanced teachers or gurus in our present society who associate themselves with a vast range of "karma" or consequences for group participants, transducing this energy to a higher value. This type of functioning, which is more than altruism, is an actual collective physiological "jump-pull" in the matrices of consciousness, and will be undertaken by those whose advancement merits such activity. Some of these advanced persons will be physicians themselves. Although such response generation now occurs on all levels of consciousness, its understanding in the makeup of familial units, communities, and larger societies can be more fully explored. The study of community medicine may therefore be a very inspiring goal for those who are reaching the age of maturity and who wish to choose their outposts in the range of knowledge.

The Medical Body

The nomenclature presently established to signify specific medical functions will require intense updating as pure consciousness becomes better understood. The naming of body parts is now based on ancient language structures that lack the power to restore meaning to the physiology. The naming of the body is an art, which by its very act causes creation to remember its mission and to develop harmonious patterns of interaction.

The body of each individual contains organs and systems which are presently named from a particular framework in consciousness. The words we use signify the perceived function of the organ, but the actual range of feeling is greatly limited. It may be effective, therefore, to rename the body with sound structures that are more useful to its function. These structures would be derived through experimentation with the light fields that create the organic derivatives of the physical body.

Those who study such structures can derive chemical flavors of consciousness which can act to identify the organ in its own field. These chemical flavors, which could be identified and actually ingested by us, would allow the body to register its own frame of reference with response to possible change. Chemists, pharmacists, and researchers in the natural therapies could identify substances which, through their vibrational resonance, would act as organic counterparts to the organs of function. This would cause these organs to restore themselves to their first memories. New waves of harmony would be created in the body through increased recognition of primary identity, allowing the organ to determine its own pathways to restructuring.

These chemical bridges could be studied through cross-technology with those who work in the area of artificial intelligence. As the body can trigger its reactions to color, sound and heat/wave activity, so can the mechanical body. Once counterparts can be made that emulate the frequency flows of the human system, mechanical study of creative inter-grounding can truly begin. The science of robotics, therefore, is to be thought of highly, not as a replacement for human evolutionary activity, but as a means of trying out new patterns of information through electrochemical frameworks and developing new biostandards of response for the human range of motion.

The development of a flexible and viable counterpart to human skin is highly desirable because it is through the skin that the intelligence situates itself in the environment. When the individual is skinless, he/she literally loses sensivity to the surroundings. The psychological sense of vulnerability that ensues is not only painful, but is conducive to a type of blindness that limits freedom and development. The encapusulation of human form in the outline of the skin creates a boundary field for deep interior investigation and discovery.

As medical personnel are able to reveal the nature of the nervous system and understand its matrical articulation with respect to the human circumference, they will be able to transfer this knowledge to an understanding of the planet as a whole. In this way, the study of human form will become the study of planetary form, and knowledge gained from the study of planetary ills will be translatable to human ills. We are offered a breathable design for the timeliness of consciousness/matter expansion. This is how globobiotic reality is made available to us. When we are more conscious, and are able to express ourselves through voluntary cognition, we will become co-blenders in the field of awakening expression. We can thaw out the neuroreceptors, creating the raw substance through which new layered existences in consciousness can take place.

This infinite drive for realization is breathtaking. Think of the graceful simplicity and magnificence it offers! We can design the entire structure of the Mother/Father God and behold ourselves even as we behold all others. As the body stretches to find its own identity, consciousness can grasp the seeds of its makeup and restore order to itself and every other thing. The body in its many houses can design the architecture for a constantly refining Universe.

Treatment Modules

The practice of housing ill individuals in hospitals may not be advantageous to the individual or to society. By treating these individuals in one large facility, the likelihood of disease contagions spreading throughout the population is great. Wellness, a condition of mind and spirit, is not easily maintained when those who are unwell are kept exclusively in each other's presence. Research into the enhancement of healing environments is underway.

The individual involved in a therapeutic program could be more easily monitored in a cheerful and uplifting environment. Far more dignity and appreciation must be given to volunteers for such projects. Their input as to the use of the information gained and its ultimate purpose must be deeply considered. Without such respect, the person who is taking part in the study feels objectified. The results themselves might therefore be affected.

When the time arrives that our entire approach to medicine focuses on education and research, there may be greater opportunities for medical personnel to act as their own study guides, performing self-experimentation which advances knowledge through personal experience and discovery. Safer treatments would be developed if physicians practiced on their own colleagues. Study would be directed more towards techniques which expand consciousness and thereby encourage health, rather than incurring the risks inherent through other forms of treatment.

In relation to the study of mental illness, we have often viewed severe difficulties of the mind as unalterable except through the use of drugs which mask or deter damaged functioning. Since the mechanics of the mind are unknown, it is assumed that by numbing the mind it can be induced to perform acceptable behavior. Once it is determined that the energy field which surrounds and permeates the physical body is a reality, it may be thoroughly studied. This will offer hope for more successful treatments for mental illness, altering this non-physical field so as to not rely exclusively on information derived solely from the physical body.

Entire structures can be designed that will simultaneously and dramatically alter the mind-fields of a collection of individuals, each of whom may have differently diagnosed forms of mental illness, but who may all be treated beneficially through one environmental approach. This approach could involve the use of high-level vibrational upliftment through the creation of subtle sound and light fields traveling through the environment and penetrating and realigning the subtle bodies of the individuals concerned. It is therefore towards the production of desirable environments, rather than ones which trap or contain the individuals needing treatment, that hospital design might evolve. These sound units, which would be structured based on advanced acoustical knowledge and the realization of its full matrical effects, could be individualized into certain compartments for more specific use by individuals with unique needs. This type of treatment would be especially beneficial for individuals who are in the first stages of life, both in the womb and during the first five to seven months of human life. At this point, the energy mechanics are far more fluid and can be benefitted greatly by energetic intervention.

Once the read-out of common physical and mental illnesses can be ascertained directly during pregnancy and early infancy, the incubator could be used for far more than simply aiding respiration. The knowledge of the field of the womb is one of Nature's great secrets which has yet to be fully revealed in our culture. The environment created in the womb is ideal to the functioning of the unborn child in all respects. The concept of floating in space and time, the stability offered both biochemically and sonically, offer the growing child in the normal situation the best possibilities for development. Science will be asked therefore to literally enter the womb in order to create treatment modules that are beneficial to full regeneration and recovery due to trauma. The consciousness is best left suspended unto itself in order for healing to occur. In this state, it has the most direct access to its own restructuring. The ideal healing environment imitates this world, which may or may not be soundless. It is the qualities that the sound induces combined with the measure of its silence that creates the possibility for regeneration.

The study of medicine in the future will be directly related to studies of the environment, both by focusing on the ecology and by designing optimum physical facilities for research, education and advancement.

The Effect of "Pollution" on the Body

When the body has advanced towards a primary level of adaptability, it is not as susceptible to being set off balance by toxicity in the environment. There are certain environments that can create optimal possibilities for the development and happiness of the individual and collective. It would be best to emphasize the creation of these optimal environments, rather than cleaning up the problems already created. Once a renewed Earth consciousness has been realized, the environment can fall into alignment with the individual and can be easily restored.

Medicine is in the position to offer the body as advisor to the elements of Nature. When Nature can work in concert with the body, a highly complementary relationship can develop, promoting many exciting and interesting results. The body can act as a window through which Nature can express Herself. In this way, the body can be greatly enhanced by the environment in which it finds itself. It can adapt to lower or higher altitudes, express itself perfectly at all temperatures, and greatly increase the sense of smell so as to easily avoid toxins.

Therefore, in the future, it may not be as beneficial for the medical profession to create chemical barriers so that the body can better fight off invasion or pollution. It would be more helpful for the body to become so adaptable that it would act as an accurate barometer spontaneously producing an internal environment favorable to its well-being. In addition, a person might find that at select intervals in the life cycle, a certain set of external circumstances might prove more advantageous. The environment could then be created to fit the needs of the individual. Forests, streams, mountains, cold or hot climates, and levels of precipitation, could all be created to develop a complementary relationship that supports the needs of consciousness.

When the body is healthy, it acts as an information conduit for the stored impulses that consciousness derives from the natural world. The body becomes an accurate reflection of the Earth as it grows into a station for advanced consciousness. Through appropriate research, the medical profession has the opportunity to enhance the biomechanical operation of the body. These enhancers would stretch the range of motion of the body, extending its sensory attributes and changing its relationship to its own center of gravity.

The doctor, as he or she becomes more sensitive to the client, can literally help the individual shape the body/Self to the requirements of occupation, inner goals, or natural movement. Relevant studies in current medicine have been done regarding adaptation to extreme conditions; for example, the situation of mountain climbers or those exploring the ocean. As medicine begins to view itself as the science of adaptation, it will bring together the studies of biodynamism and physical mechanics. This can create advanced cybernetics and bioengineering that will enhance human intelligence beyond what is presently seen as possible.

Understanding the Mechanics of Pain

A great deal of present medical research is devoted to masking the suffering imposed by physical pain. This process is viewed as a necessary and compassionate approach to human wellness. The highly advanced individual is physically incapable of experiencing pain. Pain, as we know it, is immediately retranslated into superconsciousness and is circulated as a message unit of bliss to the area needing attention. The factor of pain is presently induced by our own inability to utilize the natural signaling mechanisms of the body. In its frustration, the body can only have one response, increasing pain conductivity so as to produce some degree of attention.

When the individual is capable of monitoring the body from the inside out and can create photoreceptors that actually take pictures of areas which need solicitation and treatment, the purpose for physical pain diminishes. When something is not as greatly needed, it can naturally fade away. Along these lines, the best remedy for pain is to understand its role in the biochemical dynamics of the body and to develop different pathways through which information can travel besides that of soreness, inflammation, or trauma.

The mechanics of pain is an important area to explore. If

pain is understood as intelligence that is unable to return to its Source, we can learn how to train the body not to experience pain. Instead, a methodology of enhancement can be developed that allows the body to experience bliss at the moment that it seeks attention, sending streams of happiness, rather than suffering, to the area in need.

The individual who has mastered this level of development can deliver this knowledge to others simply by being in their presence. The body of the individual who no longer feels pain can literally teach another how this is accomplished. The secrets are traded through the avenue of mental union. This would be the primary function of a positive treatment facility. Rather than sharing in their mutual agony, individuals housed in a treatment facility would be taught methods to clear their pain, thus developing states of being in which they would be able to transcend the source of suffering.

If suffering exists at all in higher levels of consciousness, it would not be the type of suffering we presently experience. Perhaps it could be described as a vehement desire for advancement, so acute as to stretch the bonds of reality and cause the individual to shake loose from previous styles of functioning. This form of psychological suffering might continue until the individual has learned to completely identify the changing field of focus and welcome its outcome.

Historically, physical pain was used as a test or initiatory rite in cultures that valued the achievement of higher states of consciousness. One's tolerance for pain was seen as a means of stretching the boundaries of consciousness, forcing the individual to break from habitual limitations of awareness. Once the process of depth transformation can be more easily understood, the inducement of pain for spiritual advancement will no longer be necessary, proper or desirable.

The Role of the Athlete

The present system of athletic achievement focuses on competition, particularly on winning a game, with only a slight regard for the actual level of change or evolution gained by the participants. As the field of sports medicine develops, the emphasis will be on the quality of consciousness gained through physical activity. Studies can be made which monitor the inner activities of the athlete, as well as his/her ability to make selective adaptations to the environment. A highly developed union will be created between movement and consciousness.

In this context, the athlete becomes a natural experimenter in the boundaries of human activity. He or she awakens to heightened potential through a chosen sport, which opens the possibility for unlimited physical advancement. The athlete who views the body as a training station for consciousness can break through the boundaries of pain so that suffering is not equated with freedom or success. Only then can the heart open to a magnificent extent and fulfill the compassionate urgency to merge with others.

Every activity or sport can be viewed as a different possibility that can foster the melting of previous boundaries. Sports regimens would be designed on the basis of routing consciousness in ways that would be evolutionary to the individual. This would create the opportunity for the development of many new types of sports not practiced at the present time.

The athlete would naturally be involved in a great deal of self-monitoring and attention to the organic functioning of the body and would see the results of his or her performance immediately improved through such efforts. This is a very advantageous position to be in. This knowledge would extend to any craftsperson of the body who uses the physical form to its maximum in the performance of everyday activity.

Energy efficiency advancement may be seen as one of the primary occupations of the twenty-first century. Every activity has its possibility for optimal performance, and those who are interested in these areas can add greatly to the knowledge of how activity can express pure consciousness in its most lively and coordinated manifestation.

New Pharmaceuticals

The system of drug therapy provides substances which are considered specific to correct disease functions. These substances can inhibit, regulate, or destroy the primary patterning through which the disease expresses itself. This level of patterning is directly related to the progress of the disease or condition in the body. As our experience with the body becomes redefined, we are inspired to create substances which will enhance, refine, or increase the level of functioning. These substances, whether artificial or natural in their origin, are created to develop a body which will function at a reserve level of efficiency. In this manner, should invasive possibilities arrive near or attach themselves to the body, the body's defenses would already be functioning at a more than optimum level, and in a sense the body would activate its reserve defenses. Thus, any treatments which increase this reserve value are thought to be evolutionary.

One such treatment is the application of substances that increase the memory of the cells. The cell that is capable of remembering its own nature, thereby creating prototypes for itself in consciousness, leaves behind a streak in time that is like a memory map for its genetic and organic makeup. This streak may be thought of as a phase pattern in consciousness that has traces of imprinted generation from all of the previous situations the cell has encountered. The cell that is interphasic with this codal memory has a greatly enhanced chance of survival as a healthy unit, but more importantly, its future replication will be more perfect and secure. New pharmaceuticals can stress the ability of the cell to stay in phase with itself, to reproduce its own genetic map precisely, on time, and synchronistically in relationship to surrounding cells.

These new pharmaceuticals, which might be combinations of both natural and artifical substances, will emphasize this retrieval value, and will de-emphasize having to make up for genetic mutation or imperfection. Once the cell disengages its wrap and becomes more loosely defined, it lacks the balance and structure to repeat its causality in consciousness. As this happens, the cell is literally stripped of its shields and its defenses become weakened. The new pharmaceuticals will need to be primarily defensive in character, and will be linked together with increased balance in the personal and collective environment.

Internucleic Transfer

When every cell has been stablized to operate at optimum standards, it can trade nuclear material with other like-minded cells. In this position, the cells can literally act as teachers or guides, sharing information as to family of origin, related structure, and the possibility of cooperative development.

The ability of cells to feed information to each other and develop bio-synthetic links of operation is a little-known phenomenon at the present time. This option affords us the opportunity to link streams of intelligence and to thereby create new organs or structures to house the human consciousness. This is how the body will learn to remake itself. It will have good teachers through enhanced cellular knowledge, and the chain of command between pure consciousness and the organs will then have more room to reconstitute itself. As this occurs, the body will literally create new organs that will be prototypical of the adaptive needs of the entity concerned. The standard makeup of our species, which is seen as its defining characteristic, will be altered.

Human beings will realize themselves to be streams of consciousness, dialoguing with formative causality to create organic and inorganic structures. The body will then be able to complement itself by developing optimum chains of cellular command for the type of entity, its locale, and its flow in universal evolution. The body that can refine its nature and develop optimum organs of function enters into a more consonant rhythm with absolute time values. In this position, a form of barter with the relative is no longer considered necessary or desirable. The relationship of time to matter is radically changed.

The Dynamic Physical Structure

Once the individual is capable of continual and effortless adaptation, consciousness can shift to the status of being a reminder for the body, entering into a relationship that positions it well for immortality. Consciousness simply wakes up the body through constant dynamic movement; no longer called upon to restore or remake systems, the body will be able to store and replay codes of command as necessary. Consciousness will be free to play within itself in the balance of time. It becomes a ready dancer on the stage of infinite perception.

When the physical structure is bathed in such a consciousness, the body appears luminous and intense, cleansed of its static nature, and, in essence, entirely free. This quality of a body let loose to its own nature, creating itself on its own terms, lends itself to the exploration of space. This exploration, which seems mythic and perhaps impossible in our present form, will become as natural as stepping outdoors. There will come a time that those who do not travel far and wide will be considered as reclusive as those who do not presently step forward any farther than out of their own houses.

The body that is freely adaptive literally craves space. It must have room to create itself through constant renewed interaction with intergalactic intelligence. The isolation we presently experience is a direct function of a body that is uncomplementary to the union of consciousness with its full space-time value. Once we are free to travel, we will enter an interdimensional satchel of time in which the body will be laid open for enjoyment, synthesis, and increased learning.

The transitional period in which we presently find ourselves is laying the groundwork for a perfectly woven and intricately structured haven in which consciousness can reposit itself, reposition its memory, and continue onward. Our bodies will become way stations for infinite realization. A body of this type is naturally immortal, because the concept of ending cannot exist. For every cell there is a new replacement; for every aggregate of cells, there is the possibility of reorganization; for every collective shift, there is new territory in time to be bridged.

We come to Earth as a form of class reunion with beings who have streamed through our awareness in other aspects of dimensional life. The truly new experiences are few, and are often outgrowths of some other type of related event in consciousness. Newness, therefore, is rather rare for us. Without the increased possibility of originality, consciousness finds itself tearing off its wrappers, looking for new flesh to appear on the bone.

In a sense, such contracted identity can create inherent difficulties. Once consciousness is no longer satisfied with limitation, it will strongly and permanently break free. In this state, everything will be replaceable. The planet will not die, and neither will we.

CHAPTER FOUR

THE GEOMORPHIC ENERGY SHIFT

CHAPTER FOUR

THE GEOMORPHIC ENERGY SHIFT

Transductive Geomorphology

In its present state of imbalance, the Earth is dependent on the shift in time forms that will allow it to be set as a jewel in intergalactic life. The Earth is interdimensional with respect to its relationship to other mass spheres in its domain; yet it is parallel in breakfront activity only to those that are not removed from its own shielding effect. During the next thousand years, the Earth will no longer be an isolated voyager in the space-time continuum; we will move from the present sublight situation to extend our zoning to all parts of the galaxy and beyond.

As the light bodies upholding the present shielding of the Earth, we are responsible for creating the consciousness technology for implementing the shift. We cannot hold the doors open for Earth, but neither can the planet wait for us to establish our own processional boundaries. We are being forced to look out onto the field and establish a relationship to our surroundings. If we are to move into a response parameter with intergalactic life that is uniform and also effective in relationship to human life, we will have to leap efficiently into Earth time more rapidly than we might realize. We are on the verge of joining hyperspace; yet we must begin to consecrate the implied variables that will help to bring about that leap.

Friendship Matrices

Earth is composed of loops or wheels of geomagnetic material that function as light/heat/energy bands around the planet. These may be pictured as void or form spaces in which the striations depict the energetic glance of light that creates Earth's physical form. In a sense the walls that have been created around Earth are protective shielding for the unhealthy environment that we have created due to our present state of ignorance; we do not wish others to be entrapped in this snare. As we break away from the binding influences we have created and enter into free space, we will be able to journey to those parts of the galaxy that are near to us in both dimensional and energetic relationship.

Earth will not be able to sustain this contact until it shakes off its troubled inner life and transforms itself into the fully radiant body that it rightfully is. To accomplish this, Earth will have to transpose itself, energetically spiraling up its own continuum of space, light, and time, and developing its own forces of restitution that will allow it to transduce itself to its proper vibrational apex. As this occurs, there will be a shift in time, in relational value, and in our response to all of this. We will essentially be guided to the light shift by the Earth itself, and will use the planet as a wagon that will allow us to hitch ourselves to the proper point in our own individual and collective evolution.

Earth is the seed carrier for the full formation of intelligence in this region of the galaxy. We are asked to dedicate ourselves to upholding the mantle so that the planet can once again become capable of relaying the proper signal formations. Once the relay station we know as Earth has been brought fully online again, and is able to transduce the signal formations from other spheres, a new embrace of energetic union can be established. This will allow Earth to team up with those forces interested in re-establishing a sense of clear purpose in this part of the galaxy.

All of the changes we are presently seeing, both sociological, physical, and *geomorphic*, are directly related to the need for Earth to restore balance by reciting the phrase of its own epic poem, its own corollary history in what we know as time. If we choose to become significant players in this drama by allowing time to remake itself, investing in our personal evolution, we will join forces with all of the friends who have come before us. We will have the opportunity to become bearers of arms in the entire matrix of response that has been established. Should we decide not to go along, we will be given other opportunities to draw from the spring at a later date.

Our christening is in our willingness to find our appropriate dimensional pull in what is occurring and to keep progressing so that we will be able to hear the call once the proper dimensions have been set in motion. Earth is waiting for every entity situated upon it to wake up to reunite with its true identity, its true stature in the Universe. Nothing short of that will do.

The Field Structure Shift

Earth in its present form is limited to its arena of play in the substructures of consciousness. As the shift takes place, Earth will literally shed its skin and rid itself of those forces which are holding back its evolution. When this happens, Earth will cause a change in motion with respect to its spin in the evolutionary spiral. It will jump through the hoop of recorded time and enter a zone in which it is neither day nor night with respect to its own movement. From this space or void zone, Earth will essentially recreate itself.

As this shift occurs, all life forms on the planet will be asked to step aside to allow Earth to redress itself. In this shift, we will literally be called to our knees and enter a window of time in which we will be asked to remember who and what we are. All of the structures we have come to count upon, both sociological and geomorphological, will be shaken. They will be broken up so that Earth can reattach itself.

In order to facilitate this shift, Earth will enter two phases. One will be a settling of its relationships with those who presently inhabit the planet. Those who act as contributing vehicles to its evolution will be rewarded through continued life, prosperous integration, and a sense of joyous upliftment into the fold of humankind. Those who have not come to a point in their evolution in which this is possible will not be able to come aboard at this time, but will instead find themselves in the position of having to seek another home.

There will be unlimited call for those who wish to assist. It is therefore our responsibility and our destiny to try to assemble as much information and understanding as possible so that we can enter into a state of grace with the Mother's own plans for Herself and Her children. It is an opportunity to travel through the cradle of civilization without having to leave anything behind except that which is no longer necessary for our survival. All of the forms of response with respect to life and happiness must change, but with them will come other forms, other avenues which will be infinitely more exciting and powerful.

Who and What Earth Really Is

If you are a planet and you live in the matrical field of universal continuum, you are a voyager who has been given the space permit to travel the galaxy and thereby reassess your *response pools* with respect to changing modes of consciousness. Earth has been traveling in its own blanket of time for what we would consider millions of years, yet it is lonely for a time when it can truly unite with the other hosts in the array.

As Earth splits off from its present position and there is a corresponding shift in its relation to the other planetary objects, it will enter a new vault in which its hookup with the galactic continuum can be more fully established. Earth is entering a playground of the infinite in which it will be able to relive its childhood. Earth's natural structures will be returned to the clear wonder that they showed in the beginning. There will be a sense of newness and also a sense of power as Earth reclaims its own territory.

We who presently inhabit Earth in its changing form will be witness to this spectacular event and will be asked to leave behind any preconceptions of what a planet should be. We must remember that we have been dwelling on a living being, absorbing its fluids, eating its flesh, and essentially swallowing its heart for many years. As we begin to develop a more respectful relationship to the planet, we will enter a position in which our friendship with the Mother will become much more relational and transmutative.

Our sense of experiment with the planet, our ability to see Earth as a testing ground for all of the internal and external changes we wish to manifest, will afford us the opportunity of uniting with the wishes of Earth, thereby leading the way for the shift. The positive generation of changing matrices in our own consciousness will allow us to leave behind all of our present boundaries and distractions. A civilization that is light-generational does not need to rely on any rigid structures to create its *cellular memory*. As Earth reconstitutes itself, we will be able to correspond directly with its intelligence, and anything we desire that is right for our continued unfoldment will be available to us.

The present social structures, which are founded on a sense of lack rather than a sense of complete possibility, will become unnecessary. This is why we see them faltering at the present time. The degree of suffering that this seems to involve will be based on our willingness or unwillingness to join forces with Earth and unite with that which is about to happen. The interstellar plan is to restore order to our section of the galaxy and it will proceed. Two questions are posed: 1) What is our willingness to respond to the plan? and 2) Will we be able to keep up with the change by swiftly moving into cells of union with Earth itself?

Attendant Actions

Earth at this time is asking us to pay attention to its needs. It is in a state of birthing and we are being asked to be the midwives. In order to do this, we must listen to what the planet has to say about its own timing and proportion—the actual nature of the child-self that is to be born. To awaken, we must enter into portals of knowledge with those who are witnessing this change. We must act in a form of universal communion with those forces that are launching us into their own windows of space. To do this, each of us is being asked to assess what our own value system and sense of identity really are. Earth will follow as long as we pay attention to what is being asked.

This is beyond any one religious, cultural, or even scientific basis of form or reason. Even our pictures of ecological understanding are inadequate to express the coming need. Earth wishes to create an entirely new matrix of structure for its own consciousness to develop. Anything that we know as final or complete in any way is inadequate to express this. Our perception of land, sea, and air, our present relationship to all of the features that seem identifiable as planetary food, will be changed.

Earth is a living being that has been in a state of stasis regarding its point of origin; as it awakens it will reform itself entirely. It will enter a state of union with its own frame of reference and will unlock all of the formless doors that seal its present reality. In this state, it will become capable of releasing whole blocks or fields of energy that will cause amazing structural changes to its surface and to every layer of its inner core. It will remake itself to such an extent that it will literally be unrecognizable.

As this occurs, we will enter a time in which we will go along with its wishes, and we will attempt to create supporting structures for the shift to take place. Our mission will be to allow Earth to form itself and for us to stay out of its way as much as possible. Once we realize what is happening, all of the events in our own wheel will seem very small to us. We will be able to laugh them off, even if they involve incredible change or loss. To do this we must continually remember what is in store for us.

Our past, which we may regret as primitive or unworthy, is simply not relevant to this present state of affairs. Nothing that has happened on our planet is considered wrong in the scope of galactic union. We are simply being asked to assemble the possibilities through our own limited understanding.

As Earth remains constant and develops new relational pools of energy/time/heat/matter, it will guide us in the rebuilding. Until we are positioned to accomplish this, we will simply have to wait our turn and try to remember our place within the plan.

Crossing the Spectrum

Earth has been doing a corollary dance with its surrounding structures in consciousness. As we go into the temporal shift, all of the norms interposed with what we know as "down" and "up" will change. For example, if we are presently standing on a mountain and enjoying the view, as Earth changes position regarding gravity, we may feel that we are looking up even as we are looking down.

This redefined experience of motion that Earth will present will allow us to feel more advantageous about our position. We will become intergalactic gladiators, able to hold our balance by looking into the storeway of the Self rather than gaining orientation from our surroundings. We make a bridge between consciousness and matter, even as we venture outward into matter and are led inward to a realization of restored forms. Earth becomes a bridge, a repository of parallel motion in which we will be able to see ourselves standing still even as we are going forward.

As our conception of forward and backward, in and out, up and down, begins to change, we will realize that we are not what we seem. We are not crystallizations of defined cargo, the limited expanse through which we have been viewing everything around us. This realization, that we are not as solid nor as fixed in spatial orientation as we believe, will broaden our view considerably. As Earth reminds us of who and what we are, we will be able to maintain a different reaction in response to the changes we see around us.

When we are not so dependent on the established movement of Earth, we will become open to freefall so as to redefine our sense of mission, gravity, and life options. We can open the parachutes and let the reality of consciousness play itself out in all its glory within the newly opened and carefree environment. When we no longer define our humanness by our space-time orientation, we will view Earth as a marker point in the ever-changing movement of space, time and life value. This will free Earth to qualify its identity in its own way. We will be birthed out of the cradle, and the cradle will become our launching pad to the Infinite.

The Value of No-Time, No-Space

When we leave behind the sense of having to be in a certain position, our psychology can break free of a particular color, form, shape, or size. Whereas once Earth was a barrier to free-floating light/heat/matter, now preconceptions can be cast off and Earth can become a mantle from which we can contemplate our unlimited view. Earth will undergo its rotational and *parametric splits* in time with true splendor, undeterred that it is somehow creating havoc with our minds and hearts.

By freeing ourselves, we can free Earth to change. Earth will split off from its previous rotational spin, undergoing a valuation change in which it will leave behind the present space-time variables that once contained it and enter new ones. Earth will leap over the back of time and become a traveler, an adventurer, in the process of human and extraterrestrial life.

There is a split in the surface structure of presupposed time that mandated Earth's clock time and defined its nature. Whereas previously Earth was defined in terms of water, and its atmosphere created through the return of hydrogen, oxygen, and nitrogen through the surface skin of its mantle, Earth will be defined through a shift in glaciation. It will be ice-free but at the same time will be able to capture its water ratio through a different means of perception.

Earth will become capable of creating its watery surface at will, and will not be limited to the melting pots of its inner or outer core. It will be smoothly defined and shaped by its own need, desire and input. It will have its own shelves in which it will store transposed time variables and develop interlocking matrices for defining its shape, size and geomorphic stance. It will be able to call up its own numbers and relate to the geological ring of time from its own perception. It will enter into a state of grace with its immortal origins.

No planet has to die. No planet needs to be dependent upon a sun or other heat body for light/life. This dependence is a less advanced stage of life, similar to our present dependence on our sun. Earth will learn to live on its own merits. It will be able to create its own heat, cooling, and locomotion.

Every planet is waiting to become free in the vast reaches of space. Each belongs to a system or galaxy as a way station before it can advance to its universal disposition. As a planet becomes free, its purpose and nature change. When the inhabitants no longer need the planet in its present form, it can return to a form which is more suitable for its own growth and development. It does not have to accede to the plan of the inhabitants; instead, it can literally call forward inhabitants who are consonant with its own plans, its own evolutionary structure.

We will come to fully understand this living, volitional nature of Earth. We will no longer view the planet as a stripped-down version of ourselves, to be molded to our desires. We will develop a communal relationship with Earth that will be most beneficent to the environment. We will become whole visitors in Earth's evolutionary harmony.

Living in the Continuum

Earth provides a rough view for us in terms of our ability to create *void time* in interrelational space. Void time simply means the way into the Self, the seeking-out point for our own matrical structures and origins. The purpose of Earth is to provide a suitable structure for us to inherit our place in the cosmos.

Earth is a waiting chamber, offering us a comfortable ride into interstellar space. Earth holds the opportunity for us to live on the wheel of the infinite, and does this by restoring our memory of who, what, and where we are. We view ourselves as attached to something rather than simply floating through space and time. We are going to lose this orientation of living only on the surface. The Earth is an egg that is cracking open, and we will no longer distinguish between up and down with respect to the land, the sky, the sea.

It is our responsibility, therefore, to stretch ourselves in such a manner as to realize our actual temporal-spatial origins. When we are able to view Earth as a living thing that is capable of expanding our view of reality, we can enter into a consciousness where Earth fits the expanded view which we have been presented. The structures we design, the shape and size of Earth, will be created by our need to reestablish a matrix that parallels our vestigial origins, the star civilizations that seeded us.

Our time-traveling brothers and sisters of the universe see the Earth as a ribbon of light in which we as human beings are stretched out, imprinted, and catalogued. We are like data matrices to them, like ribbons of space-time that essentially at this point have nowhere to go. We are trapped in the illusion of night and day, up and down, forwards and backwards. We think that we are "someplace" when actually we are floating on end without any real ties; the Infinite is both within and without.

The illusion of the earthly momentum gives us a *crash-field identity*, a sense of disorientation in which we feel that our stable identity and surroundings are slipping away. As a result of this, we hold onto the Earth and are in a sense homogenized by it. As we loosen this grip, we will be able to stand forward and backward; we will be able to open the gates of our own reality and see that we on Earth are players in a vast, boundless continuum. In this new reality, all of the travelers learn their distancing mechanisms through freefall space intervention. The time-traveling cousins of the Infinite learn to reciprocate the laws of universal expression by subtle forms of communication. The messages sent develop chambers of activity for the evolutionary casting of new spheres of matter. We might say these chambers are like the kiln of the potter that heats the clay of finite realism to create the ceramic beauty of the finished form.

Earth is now in its rough-hewn state, limited by its need to please its inhabitants, but also aware that its final stages of liberation are upon it. As it frees itself from its earthly state and ventures into the Infinite, those of us who are coming along for the ride will be lifted into the continuum and blessed with the possibility of developing interrelational mediumship with our brothers and sisters. We are the rings of Saturn, we are the plains of Jupiter, we are all of the characteristics, both real and imagined, that make up our world. When we begin to see the planet as a museum in which we have stored the artifacts of time, we will cast off all the known objects and simply create new ones.

Every object has its purpose. It is here to signal a sense of implied mission, to provide a wholesome, clear path for us to view ourselves in our full glory and power. As we begin to leave behind every tool of self-destruction, the infinite powers hidden in these weaponry objects will be unfurled as sources of self-creation, renewal, and harmonious development. Inside every bomb or gun lies the reverse proponent. There is a structure within that pops the cork on the infinite and restores hope and balance. The force behind destruction is the same force that lies behind creation. They are really only a step away in consciousness from each other.

Although we view ourselves as seething in destructive value, we can literally change the picture by envisioning and holding in our minds and hearts a sensibility that views guns literally as plowshares for a new crop of lifeforms. The weaponry is symbolic of the full power of God as it is laid before us. It involves the fusion of dynamic powers which relate to our own consciousness. Our fear of evolution has brought us to the brink of possible annihilation, but Earth itself will not permit this. Earth will create new inhabitants who are more braced for change if we who are left here are not able to continue. There is no extinction, no annihilation, that would stop the trend of time towards a more efficacious order.

We will not destroy ourselves; we will simply remake our possibilities. As this occurs, Earth will change its structure and offer all of the technology necessary to develop light/heat power sources that are infinite, startling, and absolutely harmless. The correct registration in consciousness is all that is necessary to bring this about.

Cooperating with Earth

In order to facilitate the change, we are being asked to seek counsel from Earth itself. To do this, we must stretch our ears and allow the music of Earth to speak to us. If we listen carefully, we will hear Earth on the verge of splitting apart and creating a new system of energetic balance. We can help in this process by our own willingness to leave behind our past.

Earth is calling to us to look at new forms to discover the possible vision that is available. Earth is transmitting constant pictures of realization to us, trying to inform us of the possible ways to go. We are blind and deaf to such options because we become caught up in our own fears and the possible turn of events that seems to be gripping us. We think earthquakes, wars, and self-destruction, while Earth thinks change, new structures of formation, and unbounded awareness. Earth cannot see a war as something that is conflictual, it can only see something that will break up the options and offer the possibility of liberation. Earth cannot see a flood as destroying the soil because it knows the soil in its present form is so depleted that something must happen to develop new strands of formation to enrich it again.

Earth sees itself on a clear course of self-rejuvenation, and is asking us to assist in developing the plan. The best we can do is to create colonies of experimentation in which possible attributes of soil, climate, and environmental conditioning can be assembled. We can begin to grow new food, create new forms of fuel, and begin to envision the types of vegetation we feel would be useful to our evolution. We can literally seek out Earth by asking it what is most advantageous to its own creation.

Any new forms we make are simply way stations in this expanding consciousness; we cannot hold on to them. We must be free of the need to have a stable environment. We can grow accustomed to a continual expansion of variables so that Earth can remake itself through its own cyclical renewal project. We as humans are here to farm the fields of Earth for these changing variables, and in this process discover our true origins and abilities. Earth can offer us everything possible to magnify our sensibilities and restore a balance of perception.

The illusions that we presently live by and for can be cast off, and we can be left seeing the all in a comfortable if formidable way. Earth can provide a cushion for us until we are truly mature, truly capable of seeing that which is. This is the function of the home planet: to help us to awaken, to provide us with the tools necessary for self-exploration and renewal. Earth is not interested in holding us back or frightening us by creating a stirring of the waters. Earth is very committed to our internal wanderings in space-time, our seeking out of the God-force, our return to the full embrace of our humanness. This is the cradle of civilization that Earth wishes us to embrace.

To accomplish this, our understanding of history, of time and its relationship to Earth, must undergo a shift. We can view everything we have recorded as material to be sloughed off to eliminate the numbness of imbalance; we must move forward at a very fast pace to our full destiny. Anything short of this and we will view all of our losses, all of our sufferings, as too heavy to bear. If this is the case, we will not choose to remain at this plateau and will abandon Earth for other possibilities of development on interdimensional levels.

Those that are to be Earth voyagers will have to live in the confusing world of no-form until a resting place is found. We will be asked to stir up all of our surroundings, and to live as adventurer/warriors in a constantly expanding new universe of our own creating. The Earth will give us the tools, the machinery, the inner knowledge, and the advanced perception to carry the plan out. We have only to listen to it and to try to remain calm in a sea of worry, stress, and doubt caused by our previous ignorance as to our own nature.

When Nature can speak to us through Earth, without interruption or discontinuity, we will be able to enter into a time vault in which the planet will become our locator point in collective evolution. Earth will be stripped of all of its shielding and will be welcomed as a full galactic member. We will not remember our past because it will not exist. We will live in the now and will be restoring the time values by our continued existence.

The dissonance and confusion we presently feel is only temporary, but is understandable and necessary. As Earth changes form, so must we. This is the secret to being happy while outside the comfort zone for the next hundred years and beyond.

Earth Configuration

The substructures of Earth at the present time are receiving an outcropping of information from deep within the interior of Earth's center. Earth is essentially composed of layers of vibronic material which act as cushions to create the pillow effect necessary for the planet to store all of the information for its evolutionary movement. In this dimensional spiral, there are many avenues of existence within the planet which act as transducers for the energy as it finds its way up and down the continuum.

At the present time, the planet is undergoing a massive rebuilding of its energetic base or core. As these superstructures are rebonded they become more capable of allowing Earth to enter into its new energetic framework. Essentially, the center of the planet is becoming cooler and more solid. This response, which signals a deepening of the physical stature of Earth, will allow the planet to hold the strong glaciatic pull towards release of the surface ice structures and realigning the electromagnetic poles. As Earth retains more solidity or bulk in its mass drive, it will be able to store the necessary codework for our advancing civilization to progress.

We are literally at the center of life on Earth; our activities and constructs at the present moment directly affect the integrity and viability of the inner Earth structures. The dimensional life of Earth at its inner core, is responsible for upholding the root-centeredness of Earth's movement, and is a grounding apparatus for the essential crystallization of all metals and ores through the planet's surface. The reconstitution of these inner fields will therefore yield new combinations of alloys that will be beneficial for our planetary reconstruction.

Earth manufactures its light body mechanics by overlaying spirals of activity that translate into the surface geographical structures we presently experience. As Earth changes its scope and is capable of direct uptake of energy from its cosmic neighbors, it will become more independent and able to concentrate its own energy sources. As this happens, we as inhabitants will be able to mine from the planet substances that are more far-ranging and useful sources of energy than presently available. In order for us to receive these gifts, Earth is essentially being cleansed of certain substrata that are no longer useful and beneficial to its final stages of evolution.

We human beings who are seeking full consciousness have a responsibility to assist Earth in mining the fields of the Infinite by creating a preferred balance on Earth's surface that causes it to dive deeper into its own evolutionary flux. We can assist Earth in this process, or we can hinder it through our own misunderstanding. Even if we choose to disregard Earth's call for assistance, it is sure to find a way to shield itself from toxicity and to develop new attributes of consciousness.

What Earth Needs

Earth at the present time is in a state of incubation and preparedness. In order for it to move to its next stage of evolution, it must cast certain strata of influence into the webwork of consciousness and develop a new plan for itself. This plan calls for the inhabitants to build structures which are life-bearing to the planet, deepening the consciousness of Earth to a more profound and valuable level. These structures can be built using existing methods of construction or can literally be created by aligning with the powers of Earth that actually construct natural features.

We might say that Earth itself employs a system of energy builders. These construction units build the natural features we presently see. This same matrix of energy formation can be engendered to help us as inhabitants create structures which will be completely in the natural flow of Earth's own evolution. Such structures will literally pop up out of the ground, and will be developed for the benefit of collective units that wish to engage in planetary research and development.

The concept that Nature can assist us so directly is not a new one. We, of course, utilize the benefits of Nature on a continuing basis in our daily lives. Yet, when we bridge the gap between Nature and consciousness, the cooperative value of Nature can become much more cohesive, specific, and infinitely constructive. This is the challenge presently facing us. As we fuse our consciousness with the valuation of Nature, we are capable of growing new planetary media from that which is available. A whole host of new formations will be open to us.

Learning New Formations

The experimental phase we are presently entering involves a process of listening directly to the voice of Nature and attempting to understand its needs and demands. It is understandable that many of us feel a sense of grief, loss, and even rage at the present state of planetary formation; we have a sense of impatience about moving things along. But we should also see that the challenge to accomplish this change lies in our ability to develop positive interaction. Essentially, we are responsible for making Mother Earth our new way station to the infinite by completing projects which are uplifting and innovative.

The possibility of complete reconstruction and upliftment is not only necessary but inevitable. What are we personally to do in this process and how should our plans, dreams, and knowledge be realized? Once we have awakened to the reality that Earth is literally calling us forward and asking for our assistance, then we can develop a personal and collective plan for assistance to be fully realized. The projects we are being asked to create may involve the necessity of moving to different regions of the world from where we are accustomed to residing, and of giving up certain personal advantages in order to achieve collective goals.

The reconstruction and experimentation Earth wishes us to accomplish is in every phase of human life. When we open ourselves to the natural forces, the function of shelter and environmental foundation which our planet so lovingly provides for us will be increased manyfold. Every rock has a consciousness. Every living creature is attuned to the geomorphic pool of life that lies within and around Earth. As we grow quiet, let go of our previous identity structures, and enter into union with the planet, we will be given the opportunity to breed a new vision that is at once spectacular and strangely commonplace.

Such experimentation may require some of us to step out of our boundaries, to create colonies or way stations that are self-sufficient, innovative, and life-caring. These groupings in consciousness will hold a clear vibrational path towards our own upliftment and provide us the field of opportunity to enter into union with Earth's own consciousness.

In this transitional period, we should not feel pressured or rushed to do what we are not prepared for. As we put one foot in front of the other, we will naturally be moved to examine new people, new activities, new situations that will encourage us to try out our dreams on each other. It will be a rather slow and demanding process for us to move from an economic and social structure that is founded on a bed of fear to one that is co-creative, caring, and truly revolutionary in scope.

Each person is at a different stage of awareness with respect to his/her purpose regarding the evolution of Earth. All that is asked is that the person recognize where he/she is in the scope of evolutionary change and commit to the level that is most beneficial and comfortable. An actual leap in consciousness may be required of some, but there will always be cushions on which to land once that leap has been taken.

Enjoining Earth's Matrix

Our involvement in the upsurge of activity on and within the planet is a responsibility that is both enjoyable and challenging. As we grow into Earth valuation we can become attuned to the actual messages that our particular bioregion has stored for us. This is what is meant by Earth's encoding mechanisms. The planet speaks a language that is made up of primary syllabification that is both regional and transregional. There is a system of checks and balances that keeps the planet in a state of relative homeostasis even in its present state of imbalance. Earth communicates its knowledge factors via different elemental beings and other intelligences that are responsible for storing and defining the creative matrices of development.

As we join with these forces of Nature, we learn Earth's own language and understand what is needed where. We become farmers of Earth's own soil, and we learn to cultivate the crops that will be most useful to proper management. As we continue on this path, we learn that conscious cultivation of our land's offerings is not necessary if one can dialogue with the land and assist it in its natural process to make food and shelter.

The planet responds to our demands to create basic necessities. In our ignorance, we have created inordinate demands on the planet's energy system. We have developed a collective ego stance in which we ask Earth to meet our needs rather than assisting Earth in its own plans, its own needs. Once we come into synchronization with Earth, the planet's demands on us to work on our own behalf for shelter, economic survival, etc., will no longer be necessary.

The basis of this way of life is rooted in the inability to utilize our consciousness in alignment with Earth's purpose. As we refine our ability to identify the needs of the planet and evaluate them in correspondence to our own, the environment will become equipped to meet our needs. This can only come about by patient understanding, a willingness to create dramatic change in ourselves, and a clear realization of the picture book of reality that we are trying to achieve.

Every natural feature of Earth can be rearranged. Nothing is stationary. Earth, like our own bodies, is form-resistant and will manifest itself as necessary. When we understand that the forms of Nature are bred by our own consciousness, then we will be able to create a world that upholds the true spirit we have come here to envision.

This understanding, which appears idealistic or even dreamlike, will soon become self-evident as long as we continue to loosen our grip on our present expression of consciousness. Our civilization is now dependent on our own small will to create shelter, food, and basic economic well-being. Yet, many, many people are still unwell and starving in our midst. True freedom lies in our ability to live within the balance of Nature, and to join forces with that promise. Then there will be no one who wants for anything except continued progress and expansion. The times of survival awareness will be over.

Preparation for the Change

The best way for us to assist Earth in creating balance is to look around us and find ways to restore order. This may be as simple as planting a garden or creating other collective structures that hold higher values of consciousness. As we prepare ourselves physically, emotionally, and spiritually, Earth will become capable of making more direct contact with us, and will teach us how to merge with its own purpose.

It is beneficial for us to form interest groups for the purpose of developing new expanses of technology and consciousness. These groups might involve changes in building construction, medical information, artistic expression, diet; i.e. any shift in consciousness that will move us towards complete revivification. This is a time for planting seeds; for some, it is already a time for experimenting with different methodologies of cultivation. By utilizing the technological advances we have gained in this century and aspiring towards new ones, we will use our increased interior reception and knowledge of creation/formation; our possibilities for new structures are completely unlimited.

We tend to stop ourselves through feelings of powerlessness that limit our access to money or material resources on a vast level. Change on a small scale is also important. As we try out new forms, using Earth as a garden for change, we can leave the final arrangements to Nature. We need not plan our own wedding, but simply make the invitations, allowing Earth Mother herself to do the rest.

Every innovative cell of activity generated is incredibly pleasing to evolutionary consciousness, and is a magnificent offering to the spiral of full formation we are trying to achieve. At a time when so much seems to be failing in our present social or economic structures, it is incredibly comforting to know that simply making room allows us the opportunity to experiment, refine, and develop. We must go beyond our sense of loss and enter into new creative activity.

The shift in consciousness will manifest as a series of waves, either locally or internationally, that will challenge us to look at what we might do to offer forward movement. When the shift restabilizes, it will be tempting to go back to where we were and not continue to look for the challenge point, both personally and collectively. We are being asked to stay on top of things now and to develop change links so that we can open the way for planetary transformation. Even as the old seems more entrenched, more fixed, more resistant, we are being pushed to discover new horizons. This is the beauty and the challenge of this time.

Inferring the Spiral

In every bioregion or ecological block unit of activity there is a parallel matrix of response that acts as a guidepost for electroneural transmission from Earth's own center. Earth acts as an electronic signal outpost for cosmic energies that collect in this center and expand into the Host Intelligence or Full Field.

As Earth expands in its ability to recrystallize the matrix, a new seed formation develops that allows the planet to bridge the gap between galactic intermedial consciousness and its own staid or nonbridged formation. As this fusion in energetic activity occurs, Earth will link up in the proper vibratory docking position with other members of the light continuum. Earth can function as a true light center enjoining unit to situate pure intelligence in the proper reaches of the cosmos.

Every bioregion, every ecological block has within it beams of Intelligence, or response orders which are energetically propelled into their own grid or matrix. As Intelligence sweeps the corridors of Earth and provides the proper formations or encoding structures, these structures are situated ideally through the ecobioregions of Earth.

At the present time, however, the damage that has been wrought to our environment often prevents the proper sequencing of response into the eco-bioregions and prevents the distillation of information. Earth is being kept from its true purpose; such a blockage or weakness contributes to our inability to gain the support necessary to make leaps into the Full Field. The ecological balance can be restored through linking the mind fields with the ecological chains of command in our immediate areas of existence. Through this work an effective and powerful bond can be formed.

We are the restorers in the chain of command. It becomes our responsibility to act as relay teams for the light matrix education system to reestablish itself. In order to fulfill this responsibility, we are being asked to relocate, understand, and redefine our relationship to our immediate biosystem. This may be done in any place that seems likely on the grid of planetary heat/light formation. As we become sensitized to our immediate environment, we can rediscover the centers of possibility in our own consciousness.

Earth in its living responses continues to signal its formation system to us. With growth in interest, sensitivity, and conductive force, we can begin to construct houses, gardens, and energy centers so that more complete transmission and conductivity can be achieved. In this transitional period, the eco-building blocks consist of those individuals and groups who are willing to stand at the front of Earth and listen carefully for its own series of instructions. As Earth begins to signal its response formation, it can provide many possible blueprints of awareness; we will recrystallize the matrices and develop a chain reaction that will assist in the shift in time and consciousness.

Earth is a guardian center for activity in this sphere. Its responsibility to the entire eco-region in this quadrant of our galaxy is clear. We are being asked to help restore order by becoming listening ears for Earth. This can begin as simply and as literally as standing in our own backyard and acquainting ourselves with the impulses of Nature that are crystallized therein.

Although it is certainly true that some areas of Earth are more powerful in their response mechanisms, there is not a spot on Earth, not a stone that is placed here, that does not have a true purpose or responsibility in the plan. Everything and everybody is important. As we begin to awaken to the thought presence of our nearby locale and attune to its essence, the nature of its structure, our sensibility and definition with response to Earth will become clearer and more effective. This is the challenge we are facing now.

Nuclear Power

The alarms that are being sounded concerning the proliferation of nuclear-powered generators and weaponry are merited. However, it should be understood that the very same nuclear response mechanisms that have been created in our sphere can be used as output stations in the communications network as it is established.

The nuclear power station presents a high degree of toxicity to the environment. It is, however a bridge between the understanding of atomic synergistic fusion and the present means of polarization developed for the purpose of splitting or reducing the atom. The concept of atomic replacement or the ability of the atom through its own nucleic bonding mechanisms to create new forms, new shapes in consciousness, is essential in the field of higher consciousness. As we begin to break from our past time codes and understand the value of nuclear dynamics we will be less afraid of the power of the atom and more able to cooperate with it.

The atomic center in our nuclear mandate involves our willingness to utilize the present radiation centers as transformers for the infusion of pure embryonic energy into our sphere. We can literally birth new pioneers, new ecospirits in consciousness, via the nuclear power plants themselves. Though these plants now house deadly plutonium and other elements that are rightly deemed dangerous or toxic, these same elements, treated in a different framework, can become useful members of the team of isotopic life.

Our ability to share the ingredients of consciousness with the radiant elements is our responsibility and our destiny. If we choose to utilize nuclear energy we must gain the ability to control its toxic response mechanisms. This will enable us to make not only rocket fuels but also chain-reaction matrices of great proportions that will positively influence the environment in an area, without the destructiveness of a nuclear detonation.

During nuclear detonation, the energy formations are essentially relative and planar in their responses. The area becomes filled with toxic or burning material that can destroy the inhabitants and their genetic possibilities for many generations to come. Nuclear power plants can be seen as possible vehicles for pure energetic encounter; the energy formations therein can be recrystallized at different epicenters of consciousness. Then we will be able to draw a ring around the moon and create nuclear relay stations that will act as matter contact points for our present civilization's encounter with other life forms.

Atomic fusion is a powerful alternative in that it allows the atom to release its internal power without creating harmful effects to the environment. In looking towards fusion reactors, the knowledge of the internal expression of consciousness hidden in the atom can be revealed. Our aim must be to unify our wider scope of vision with the power inherent in the atom. To do this we need to view the atom as an information gateway—a matrix formation of consciousness that can reveal the expanded Universe. The atom offers us a means of in-depth communication with the mechanics of Creation and thereby with that which God has created. Finding peaceful use of atomic energy demands heartfelt respect and cognition of the balance of power in all of Nature.

The atom, like the Earth itself, is a communications chip, a synergistic passageway into the most profound levels of information gathering. This information is for the purpose of spiritual upliftment and transformation; it is the "fiber optics" of the Universal Light, the codes of intelligence necessary to enter God's plan for our world. The interplay gained through understanding the flow of atomic energy will allow us to gain access to an explosion of the soul, an explosion of Self discovering Self.

Geomorphic Resonance Fields

Earth is composed of photo-resonance chambers for the absorption of light/heat/energy into certain parametric layers. Each of these chambers acts as a superstructure that produces the kind of *bioreticulative* energy patterns that cool the planet more rapidly and recrystallize new formations. The planet is resynthesizing, moving into chains of activity in which operation can become more constant and creative. Earth will then be able to spin on a different rotation than its present one and will develop a different gradation of space-time intervals. As the planet embraces the time shift, Earth will rearrange its time spirals so that it can line up with the new uptake of energy available. Earth will develop chain-reaction spirals that will uptake energy directly into the matrices or front intervals for warming or cooling.

The temperature dynamics of the interior of Earth are very important right now as Earth develops new chambers for cooling and heating. As the planet undergoes a frame shift in its reference points, it will become superfluid and more capable of leaving behind trace materials that are no longer relevant to its smooth skin. The planet is undergoing a refinement process of immense proportions. We can contribute to this process by underlaying the matrix with the proper formations of light/heat/sound transfer. We can support Earth by allowing it to cool evenly, developing the proper attitude about the planetary infrastructure. By allowing Earth to come into its own formation and not attempting to restore what we believe is proper, the path will be cleared for new vegetation, new life forms, and the possibility of an exciting future.

Although we rightfully feel a sense of mourning when we see our planet as dying at the present time, it is also heading towards rebirth—a new lifeform, a new energy matrix of startling clarity and color. We will have before us a planet whose scope of awakening majesty will be breathtaking. The dying-out process that is occurring now is due in part to environmental pollution. Some species will no longer be present; some will literally be flown off planet. Yet, the disharmonious conditions that are depleting our spirit and energy will eventually give way to incredible flowering and upliftment for our planet and its inhabitants.

We who are part of this shift must learn a great deal of patience, understanding, and a willingness to be of service. Although it may seem that we are being left to our own devices, we are actually being brought along by forces beyond our comprehension. We must listen to the heartbeat of the planet, responding to the vectorial compromises that the planet is making to mathematically correct and refine its present structure.

Understanding Matrix Formations

The vectorial analysis of Earth at present leads us to understand that the synthesis of Earth time and cosmic time is about to unfold. In this synthesis Earth will spin differently in relationship to the cosmic clock and will move forward in its structure/form base analysis so that it can merge with its new parameters. In this corridor, we are like birds sitting on the branch of a tree waiting to be shaken off or clinging tightly to our branch. Earth is preparing for a climactic intervention in which the shift in emphasis within its own dynamic structure will allow it to play more precisely within the intervals of time. Instead of holding its own in a field of change, it will bounce forward into the infinite structure, redefining its position with respect to time itself. In this interplay, Earth will become a fantastic voyager in the interdynamic wheel of life in our galactic zone.

The parametrics of the Earth change are based on the simultaneous process of vibrational and metamorphic shift. In other words, Earth is a compound of energy, form, and vibrational upliftment. As Earth moves into the zone of temporal majesty in which it can decide its own time parameters, it will gracefully enter infinite perception. The planet will shed its skin and develop a holding pattern which will support its move into infinite realization.

This is a birthing period for our beloved planet, and we as its inhabitants can function as the mothers. We are being asked to loosen our hold on our former environment and enter into a void time in which the outcome is suspended. As the corrections in motion take place with relationship to Earth's axis and its rotational spin, there will be changes in weather, geographical activity, and behavior on our plane. These changes will be fostered by a need to return to our primary source of attention: the planet itself.

Once we tend to our home and develop life jackets so that we can swim in the sea of the unknown, we can then come ashore and begin to build anew. Earth is allotting this portion of time as a cleanup period and is inviting us to pick up a broom!

Vectorial Squares

The infinite correlation of Earth time with cosmic time demands a form of arithmetic breakfront that splits reality into complex squares of evaluational memory. We might think of them as circuit boards stretched over a wide field of space-time. The designers of these boards are the patterning structures of time itself. As time is stretched out over Earth, the vectorial crossbeams of collective consciousness will be smoothed, shaped, and recolored. We will experience sweeps of time in which the vectorial function mediums will be introduced to reshape the entire structure of the planet.

As these vectors of comprehension are developed, we as inhabitants will also be stretched and asked to re-examine our psychological constructs of morning, noon, and night. We may live in a temporal timeless zone in which our sense of historic development is indistinct. In this resting period before maximized procreation we will be able to cradle Earth in our arms and develop a form/feeling relationship to it. We will prepare ourselves by opening to Earth's movement, reliving our experience of planetary formation, and becoming brothers and sisters to our awakening planet.

Earth will forge new vectors of comprehension with its own relatives, the nearby galactic spheres that signal homecoming. As these circuitry boards are put in place, we will be able to travel to the stars simply by entering the chambers of magnetic resonance within Earth's own construct. This will provide us galactic travel passes.

When the vectors of comprehension spread themselves, and Earth is able to loosen its hold on the parliamentary formations that are its present timetable, the entire structure of Earth will shift. It may seem as if it is happening all at once, but of course the buildup is now occurring gradually. As the entire matrical structure fuses with itself, and the formations of new Earth matter are complete, we can return to our everyday business from a whole different order of perception.

Returning Earth to its origins does not mean going backwards; there is no forward, no up or down in this nontemporal scheme. We will navigate by feeling under the surface, exploring our vantage point from many different altars of consciousness. We will step inside ourselves and see how we are made.

Exploring the Landing Gear

Earth is aligning itself to the ventral chambers of awareness locked within its surface structure. As this is accomplished, our land shoes will be given to us. We can venture into a space-time dimension in which we can live in zero gravity, feeling weightless while at the same time free to move about without fear of falling free into space no-time. A sense of liberation is achieved by resistance to the power of gravity. When we no longer need this force to hold us together, we will be able to split apart into cosmic wavelengths of geodimensional matter in which we will not be chained to any fixed perception. In this state, the landing gear will be our own adhesive consciousness, gluing us together through the constraints of implied form.

The formless human being, at once free and also upholding the structure of changing values, is offered the opportunity to experience interdimensionsal mind-meld with its true origins. When this occurs, Earth will rise to the occasion, developing the physical and vibrational structures necessary to support this nascent life. The exciting part about all of this is that it is happening to us, and our job at present is simply to not resist, to develop an inner relationship with ourselves that will create a smooth shift, a smooth causality of being.

Earth is providing us with the wheels with which to brake. We have to trust that breaking the necessary boundaries will be less painful and more joyful. Earth is offering us a passport to motion; we need to permit the galactic structures to bring us on board. Our sense of humanity, our sense of thereness, rides in this balance.

Our ability to stretch ourselves through the parallel gates of space-time is the beginning, not the end of this voyage. We are looking into the future through the gates of the past. All of the wars fought, the disasters met, might seem like a blink of the eye in the cosmic calendar. We can look into the eye of the storm and create windows of opportunity in which the receiving patterns, the matrical interlays can change. In this vision, our inner nature can line up with its own self-mastery, its own continuity.

The vectorial transition, the re-blueprinting of infinite structure, lies within our grasp. Interrelational biodynamics demands a vast field of experimentation and observation of possible relief forms in our continental divide. It is up to us to discover them.

Activity Functions

Earth is in a state of transitional symbiosis; as it meets up with its ancestral twin in time and space, it will fuse together with its biorotational orbit and become bonded in order to reproduce. In this sense, Earth will be able to create new bonding material, new planetary structures from its own surface. The formation of new planets or orbital spheres in our region is essential to the development of other chains of command and to assist in the responsibilities that will be generated.

Earth's mating with itself and redefining its parameters will create the vehicle necessary for biorotational change as it is presently mapped out. Earth is systematized into quadrants of activity, each of which has a distinct function in the biorotational scheme. When Earth fuses together with its essential ordering compound, it enters a free-fall state in which it can link up to its orbital twin and reestablish itself.

We can think of this as similar to the embryonic wedding of genetic material that brings together two people from our species. Earth teams up with its new respondents in the field; it can develop different pools of origin, different rotational spirals in its metaform structure. As this occurs, Earth will supercool to its twin, merge with it, and form new ice shields which will re-establish genetic grids throughout this part of the solar system.

All of the life forms presently residing within our sphere depend on this speeding up of the mutational spiral in order to bring new life forms, new points of origin into our zone. As this accomplishment is cleared, the genetic relay system of possible pools of light/heat/wave activity can be processed, speeded up, and redefined. Earth will become a brother/sister of itself, sharing its mutational values with other species of origin.

The sense of a living planet reproducing its own ice crystallization along the formation of space-time is not a new one. Our ancestors understood the causative value of Earth. They knew our planet to be a living, breathing, biosystem capable of creating its own children in the scheme of biorelational life. As we coat Earth with our own knowledge, our own field consciousness, it will expand, shift, and stretch its present parameters to create new possibilities of genetic life.

Earth holds the memory of all its countertypes, all its productive forms. As the planet mutates, rediscovering its own process in awakening consciousness, it will be reintroduced to the scheme of life. Thus Earth will marry its true origins and becoming capable of superimposing its own shielding process in the direction of other lifeforms and points of origin in the Universe.

Earth is essentially the mapping station for another like planet to re-intersect with its own value. As Earth is speeded up it will merge with this point of origin and re-establish its true identity or cultural/racial stream.

We may think of Earth as a castoff, a breakaway pattern from the pure value of consciousness intercedent in this zone. In its renegade state, Earth has become capable of developing chains of being that are not temporal to this region; they are stricken from it and left alone to fend for themselves. We may think of Earth as an experiment in intergalactic intervention, a means of determining the likelihood of joining together a vast array of different but understandably related species and merging them together so that they will come to know their true origins.

As we return to our central cast and understand our differences, we become capable of looking at the reason behind the chaos and developing a new chain of command that will bring a great sense of hope and prosperity to us. The differences in our biorelational origins can cause animosity towards each other but they can also catalyze a good-as-all mentality that promotes fusion with our God, our Creator.

Every difference that we see, which is not imaginary or simply through the experience of form, represents a striation in the majesty of consciousness itself; it is built into the picture of earthly life as a metaphor, a corollary of consciousness, for us to weave new biorelational threads, understanding our interposal relationship to the cosmic whole. By envisioning ourselves reuniting with the one Source, we can develop an actuarial blueprint of awakened life that reasserts our temporal origins but also stretches us onto the template of biomagnetic experience.

Earth is becoming *radionically* separate even as it forges a new identity with its sacred twin. It is becoming capable of mooring to the distant star regions, developing a full synchronistic value with its heart-self, as well as its nearest neighbors. Earth will develop this radionic influence through our role as inhabitants, as voyagers on the surface of reality; we will enter into the same magical symbiosis of space-time that our own planet is experiencing. We will have the opportunity to cross the lines of reason between what we know as life and death, day and night, all of the opposites of relational value that we now take for granted. Earth is our teacher; our formative vision lies within and upon the planet itself.

Living in Communion

Earth can develop its own relational field of influence only as long as we unite to help the process along. We can assist by becoming energetically connected to the planetary interface, and developing chains of command whereby we can communicate directly with the different states of reaction formation in which Earth is presently engaged. As we mate on the eve of night and wed together the streaming influences that create biorelational synchronicity, the power of the light and dark channels opens up to us and we literally can make night into day.

The whole atmosphere of being bio-related to the planetary formation can change. The feeling of being strung out at the end of consciousness and left to wrap around our values, strangled by their twisting around us, will loosen up. We will begin to perceive Earth as holding us to nothing, but instead as being a miraculous band of interstellar light around our awareness. We will be infused, drenched to the skin with the sunlight of awakening form. The power of our cosmic birthright will become self-evident to us.

To be human is to share in the God plan for a cosmic Earth; free and unchanging in one sense, yet stretched infinitely into a myriad of changing forms and possibilities. Once we receive Earth and understand its mission, we can enter into a state of harmonious union with it in which we will act as the enchaining mechanism to encircle and redefine the entire vision of human functioning.

Earth is modeling itself after sister planets and stars which share a similar vision, a similar outpost in the cosmic career. Once we learn how we are interrelated, how we have been formed, and where we are going, our sense of a living structure will become more powerful and evolutionary. The planet demands a team approach because this is the only way that we can link to our star formations and create the necessary influence to rebirth ourselves to our own identity.

This process is the Living Earth manual, the recrystallization of genetic material in our substructure and the releasing of a true formative vision in our perception of holistic reality. The visionary gateway is through a biocentered, radionically balanced planetary system. The structures are now inexorably joined with the holding formations that link us to the Universal Light. As we create the junction points for this accomplishment, Earth will leave behind its old form, burst the cocoon of its present stationary state and open itself to perfected vision and accomplishment. From this point, all evolutionary structures will be remodeled, reshaped, and transformed into the basic building blocks of DNA, which birth new lifeforms.

It is a time of holding together with the Infinite, sharing in the possibility of witnessing the creation story from our own seat in awakened vision. Unlike the fish or the bird, our knowledge of life, our understanding from a megaconceptual vantage point, places us in a unique position to bring about and midwife the new creation. It is an exciting though challenging responsibility.

Upping the Periscope

As our vision rises to its origins, our reality becomes more clear, and Earth will feed us conduits of *interlitic* material that will build the essential foundation of our new form. Earth is starlight-shaped; in its relative state it is synchronous with the star formations that have sculpted it. When we enter this star pool, we develop relationships with our brothers and sisters in the continuum. Our manifest destiny would be to link to the nearest neighbors and establish relay stations to receive and transmit valuable data about the formation of consciousness in our zone. As this is accomplished, our friends will learn the secrets of our own unique symbiosis and will attempt to use them.

Our willingness to bridge the gap between our isotopic bonding fields and our limited viewpoint in *sequestral* matter will open us to share knowledge with the star kingdoms that await union with us. It is like looking out into the sky and seeing the Host from its most magnificent vantage point and understanding what will occur by such a union. This sense of accomplishment will help Earth to become a registered medium in the interstellar cookbook. Our experience of restitution, planetary transmutation, and birthing of the new beginning will be complete.

As fast as our minds and hearts can stir things along, this is how fast Earth will progress. It is a team sport involving awakened vision. Our interior perception becomes more uniform and conscious; the Host arrives on the scene to straighten us out, developing interrelational vision with us as was done at the begininning of our Creation. Our attempt to relink with God more directly is an essential feature of this plan. The skill of communicating with Host Intelligence will reveal to us our purpose, our destiny, and our true origins. As our Creator speaks more clearly to us, we will be able to speak more clearly to ourselves.

Entering this field of continuity, the essential forces of Nature will streamline themselves, curtailing their activities so that the entire surface structure of Earth can become smoother, more habitable, and engenic of awakening life. The forces of nature and of our own awakened consciousness will become a melting pot for this crystallization, this valuation to occur.

Our perception of the full extent of the plan is limited only by our inability to grasp the identity matrix of our planet. The planet becomes a restless field in which creation/manifestation is brought about and reinstituted in its pure form. The Heaven on Earth value thus becomes fully realized in the *upscope* of awakening vision.

Fashioning the Earth

Our present relationship to Earth as a place to develop a range of advanced technology is admirable, although it limits our ability to define and understand the mechanics for our species' advancement. Earth has its attention trained on the sky and yet we continue to try to hurt ourselves through developing technological values that do not uphold the true integrity, the true spirit of the planet. Our primary responsibility at this time is to place attention on the development of clean technology that respects Earth's desire to move towards a renewed sense of expansion.

On practical levels there is much work to be done and we are very capable of carrying it out. Some of the necessary changes include the recycling of solid wastes, clearing the planet's waterways of pollution, and introducing positive reaction chemicals that will essentially scour the air of pollutants. We can aggressively pursue a cleanup campaign for Earth's environment while continuing to explore our understanding of advanced mathematics, chemistry and biology.

Earth has a wake-up force that is calling to us now to come together to reduce the deleterious effects of many years of neglect. Since Earth's ability for self-healing is very great, a start in the right direction is all that is necessary now. Once we give Earth a boost through our understanding and efforts, the planet itself will be able to take over and develop new systems of technology. We are being asked to set our compass in the right direction and to make the integrity of our environment the most important issue in our awareness.

Landfill systems of waste management that do not include replenishing the Earth's nutrients are not a beneficial approach to the present environmental crisis. Considered more useful are massive waste treatment and recycling facilities that utilize positive combatants to neutralize the toxic effect of hazardous waste materials and then reintroduce them comfortably back to the land. Earth has many holding stations of its own in which waste may be safely reintroduced. By exploring Earth's own reservoirs of consciousness, scientists can gain access to how the planet actually refines and develops its cleansing system. They can then work in conjunction with this system to develop strategies of action to help bring the planet back into balance.

The problem of nuclear waste in particular is seen as especially grave since the pollutants involved must be completely neutralized in order to be discarded safely. This may be accomplished through the use of anti-nuclear matter. This technology demands a more advanced knowledge of atomic structure than we have at present; yet the ability to work in this direction for positive ends will be supported if our reason is right. For example, if our motivation is to learn to neutralize radioactive material more effectively in order to wage nuclear war more advantageously, the technology will not be shared with us.

Although we seem to be progressing quite rapidly in the field of technology, our access codes to certain systems of understanding are extremely limited due to our inability to create a state of peaceful coexistence with our neighbors in the world. As we move into union with our neighbors and put pressure on each other to conform to standards of tolerance, decency, and cooperation, our ability to solve our problems on a global level will greatly increase. The choice appears to be ours.

Entering the Neutral Zone

This interval in our relationship with Earth may be seen as a proving ground to signify our good intentions. If we are ever going to transform the planet within the next twenty-five years, it is essential that we change our planetary awareness and attitude. Every problem has its antidote, and our ability to stretch the limits of consciousness is entirely unlimited. Yet we cannot go forward unless we are willing to take responsibility for the mistakes we have made and create a complete dedication to their solution. Any small effort in this direction is important, although the push is on to make all the smaller efforts merge together under one umbrella.

Our tendency as a race to live separatively and to have many groups committed to similar causes competing for limited financial resources and publicity is counterproductive. It would be more beneficial to develop an extensive worldwide commission to deal with our present environmental crisis and to leap into the twenty-first century with our goals and ambitions firmly established. To stabilize the environment, a system of positive interventions is in order. We can treat and neutralize all of the major irritants that contribute to pollution, and we can develop new streams of technology that are innovative and neutral in their effect on the environment. The truly enhancive technology which will rebuild the planet will be introduced once we are completely committed to living in the neutral zone for a period of time.

We must make a commitment to stop where we are and to live in a transitional value for some time until it is seen that we can behave ourselves. At that point, Earth will open itself to our desires and there will be a great increase in our ability to develop innovative plans. This twenty-five year period is seen as a positive turning point. We will move forward as necessary, even if there are people whose perspective creates obstacles to this path. There are certainly those among us now who do not wish to make our environment, and indeed our entire relationship to evolution, a priority in their awareness.

In order to work with this reality, it is suggested that those of us who understand the necessity of these times come together in colonies or environmental way stations to support each other and work for common goals. Those who do not wish to come on board at the present time cannot and should not be coerced to do so. We need to become leaders in creating intercessional technology that will strike the balance. It is not that we should ignore the naysayers in this effort, but we should not engage in actions that tantalize the opposition and cause these people to be galvanized against our aims.

The positive course of action is to acknowledge our predicament and create a strong measure of support that will unlock the necessary keys to turn our situation around. Practical suggestions in this area are very numerous, and really come down to one area: cooperation. If we can learn to live more harmoniously with each other and develop a more sharing attitude, we will be able to feed, clothe, and psychologically nourish each other through this difficult twenty-five year period. If we cannot, it will be a time of isolation and fear.

All of the forces that support Earth and are working towards the planet's benefit, on every level, from the human to the divine, wish for this time to be as peaceful and transformative as possible. The strategies in this direction may vary but they all work towards one aim: planetary upliftment and a return to stability. This is being sought by everyone who understands the situation and any projects which work in this direction are considered admirable.

Cooperative Projects

The areas in which transitional technology must fall are medicine, law, commerce, and psychology; in short, all of the principal areas of human interaction and spirit. Changes in how we view what it means to be human must go hand in hand with the development of new standards of communication and industry. The development of advanced systems of communication is a high priority, since once everyone can link together it is less likely that we will be able to unite around destructive aims.

Other areas of emphasis include the development of programs to plan the use of artistic resources to illustrate world possibilities. Every seasoned performer is encouraged to think about how his or her artistic sensibility might contribute to the development of new standards of beauty and achievement. Our collective imagination can be sparked without losing our inherent individuality and humor. This will make it possible to bring about a dramatic rise in consciousness for the benefit of the planet.

Developmental Environments

Three quarters of the present surface area of our planet is composed of water. Our main uses of water are drinking, recreation, and transportation. Water has limited value as an avenue of transportation due to our decreased dependence on travel by ship. We also utilize water to lay down communication systems and to dispose of our wastes.

The richness of the oceans offers significant possibility for exploration and development. Our ability to utilize the oceans as way stations for experimental research and exploration is quite important during this coming time interval. All of the research that has been done up to now focused on sustaining ourselves underwater via mechanical means. Emphasis has been on the development of advanced diving apparatus and submarine vehicles. These alternatives create a type of protective screening between ourselves and the aquatic environment. It would be more evolutionary for us to consider merging with the environment through the development of genetically evolved systems of operation which, when combined with research into higher states of consciousness, would enable us to exist for longer periods of time in amphibious circumstances..

Our ability to live beneath the surface of the ocean and to function there independently will allow us to explore whole new realms of agriculture, architectural design, and commerce to be explored. When water is no longer an alien environment to us, we will see the entire Earth as our playground and will be able to travel readily to any part of the planet. No geographic or climatic feature need be an obstacle to our rapport with the environment.

Land Aquatics

The amphibious nature of human beings can afford us the option of developing interrelational values between our waking and sleeping consciousness. In the medium of water, there is a natural return to a womblike state of brainwave functioning that is conducive to understanding higher potentials of creativity. When brainwaves are slowed down in the aquatic medium, a shift takes place that allows the individual to enter a dreamlike state in which there is a greater possibility of leaving this dimension of reality. We can begin to sift the unconscious more thoroughly and enter into a union with our paradimensional life.

Aquatic environments provide a slowing of aging and the opportunity of learning to develop more stress-free surroundings. Intelligence that is water-based can smoothly coast along the sea of challenge, developing an interlocking value structure with the realm of ancestral knowledge. Once stripped of land-based functioning, the individual can become more sensitive to other existences in his or her own consciousness, and can learn more easily through mirroring certain "portent" memories.

The physical musculature can remain in a greater state of health for much longer through water exercise, but this immersion must occur in water that has not been altered through toxic substances, whether through pollution or the purification systems of swimming pools. The natural currents found in water stimulate and promote regeneration in the nervous system, and can effectively treat symptoms such as paralysis or entropy in the entire biological system. The type of telepathic breakfront communication that will be developed in our species is more readily facilitated in an underwater environment. The breaking of the boundaries in an aquatic environment links us with other species more efficiently. We learn to "language" ourselves into interspecial domains of awareness. Our insistence on species separation and superiority is one of the attitudinal blocks that can be altered so that we can experience ourselves as interlinked with the formative intelligence of Earth.

Life Integers

As Earth steps up its vibrational spin and reunifies with its essential nature, we as land-based creatures can experience every domain of our planet as home. When air, sea, and land are equally accessible to us, we can utilize the mechanics of *vibro-dimensional ribbing* to find the striation variables that can develop new waves of *intercedance* in our evolutionary makeup. We can open to new clothes like new leaves, experiencing our adaptable body/mind formations as acceptable modes of transportation and creative function.

The oceanic planes create sonar pools for diving into unknown ports of awareness where lies revealed the layered superstructure of our world. Gradations of refinement between the *coast mediums* and the intercessional units of space-time that make up the planetary interface can cause us to leave behind our inherent fears of clock-dimensional change. We can then discover a time-bending spiral of mind-webbing in which we coast through the environmental checkpoints unperturbed by psychological heat or noise. Frictionless, open to the pool gates as the spin of the planetary hum moves us along, our freedom to develop an entirely new relationship to the meaning of life awaits us.

As we realize our universe in purely mathematical terms, we can internally hear and see the mathematical relationship between every aspect of the environment and its equating value within us. We enter into a network of pure cognition in which learning becomes instantly developmental and wholly creative. With Earth as our interplanetary medium, we can coast between the worlds of sound, color, and light, digest their fragrance, their rhythm, and develop spirals of relationship with them that identify our sense barriers and restore order to our inner and outer environment.

The trans-reaches of our own inner kingdom rest with our ability to realize our sense drives as directional media in the upscope of paradimensional life. The Earth can relay any message we seek, and can leave us breathless in the search for new opportunity. Every routine of the mind must be resaturated to accomplish this.

Dimensional Mind Pools

Earth switches itself through a constellation of mental reaches that circumvent stored memories. The planet reroutes them into more advantageous norms of response. When it develops this switching mechanism in such a way as to enable consciousness to place itself squarely on the back of change, Earth will rearrange itself, and the geography of continuity will become the geography of perpetuity.

The infinite value of "I–Thou" relationships causes us to view our environment as enveloping us but distanced from the personal "I." This will change as we view the environment as a coastal extension of the personality; the "I–We" relationship surfaces, and the individual becomes able to interface as a fully engaged force with his/her lifestream. The new vantage point combines the sameness of continuity with individualized breakfronts; the normalized dimensional pools stretch the distance between separate and unified forms. The resulting structure is inter-relational, organically solid, but fluidly motionless. It can open into the tongue of the Infinite, and hold its own there.

As the individual lapses into centrality with his/her normalized vision, the entire set of parameters which are environmental in scope collapse back onto themselves. The individual becomes vested in individuality even as he/she is reeling in the apparent. The proximity of causal transmission becomes clear. He/she wakes up to the uniform, establishes a base of operations there, and can realize the vested structure of consciousness.

The mind becomes bound while at the same time completely open to its inherent channels. The unbound mind that is wrapped up in the uniform is by its very nature free. Since Earth is not really holding it, but is melting the forces of separation, new understandings are crystallized creating medial entryways into developing form; the individual is held to Earth by his/her causal imagination. We have a foothold in the finite, but are unbound, free, and contiguous with our paradimensional scope of activity. The individual can then become a freely intelligent and highly equipped inhabitant of the plane in which he/she resides.

The valuation of down or up, in or out is inherently meaningless; the evolutionary framework, the jump-working of the cognitive mind, becomes completely intercapsulary and open to its own investigation and discovery. With this stance in place, the individual is unified with Earth, bound to it as a mother-home sequence, but ready to spring forward at any moment into his/her own causality. That internal readiness provides the landing apparatus in the free-fall.

This sequestered state is a meeting ground for the imagination. It links the individual to Earth-home for the purpose of insight, interchange, and apparent survival but it does not limit him or her. The individual leaves behind striations of awareness from previous sets of events; personal life is left behind. The home infrastructure is now composed of sequences of random circumstantial happenings strung together for the amusement and fruition of the individual and his/her collective perception. Life enters a state of apparent tranquility, even as it is dynamically stirred at its core.

The miracle of life-centered activity is that it leaves room for a sense of home or security even as it breaks the boundaries of conscious standards of form. We become grounded or centered in a rooted formation with Earth's own pool of union; at the same time we are completely independent, open, alive to the value of the "We." This is the stretch that individuals make in the leap to full vibrational continuity and fluid perception. Open to the dominant stream of affective cognition, we can leap ahead. Thus we can take Earth a step beyond its own free and encompassing matrix so that the parallel of the uniform can be achieved.

Free to enjoy Earth's magic in awakened activity, the boundariless twilight of the uniform is revealed to us. All of the cognitive and seemingly dissonant structures that are available shift back to this one continuous and harmonious formless form. We share the restitution of the planet with the restitution of ourselves. From the sequestral vantage point at which we find ourselves, there is no difference between them. This is the place from which we are looking—a peephole on the infinite.

Dimensional Effects

As Earth steps onto the field of fluid cognition and can recognize itself in the mask of infinite reality, the faces that it makes, the interior dynamics of change are realized. This opens the door for us to identify our purpose more precisely. We are engaged in a streaming effect with Earth, in which the systems of intelligence that lie dormant within the planet are laid bare to us. In this grid of structural cognition, all of the lampposts that have been references for us, our locator points in the field of stripes and squares, are cast off.

Yet, Earth has a job to do. It wishes to strike out on its own, much as we do. In this sense, we can loop back around Earth, secure our flag to the pole, and develop a continual objectivity that will inspire us to share the secrets of the planet without having to possess it. We can reach ourselves and Earth at the same time. There is no opposite value here.

This unification or stretch value will allow us to slingshot away from the spaces of tongued perception, the footholds that left us grasping for continuity; all of our friends will be stationary objects beyond and through which we are circling. Our relationships to others can then be viewed as streamed events in the space-time map—experiences that unify and ground us but continue to demand that we be reaped by change. We experience ourselves in the different objects of perception inherent in our field, even as a perpetual, evolutionary force pushes us along. Our relationships—with ourselves as well as with those of other species, other apparatuses of conduct in the universal cognition—will be viewed as opportunities for parapsychological advancement; we will find that we are in the optimum position for the type of cognition we seek.

The evolution of wakened form involves a stretch of imagination, but not the imagination of Psyche bound to form. It is the imagination of the interior makings of consiousness, a mind recreating footprints leading back to Self in the sands of time. This type of imagination is a creative process, open to Earth's own lettering, magnified by Earth's value of perception/realization. It is a uniquely human activity because it involves our planetary actions, our planetary sensibility, our feeling of home.

The Earth functions as an imaginative corridor through which time can be stretched and redefined. The blockprints we leave, the steps in rarified vantage point that are expressed, recrystallize the kingdom. They make it possible for us to be comforted by the new species even as we are making it. We can jump into the future-time hold, and leave only the present doorways unopened. As we open the future gates, the present is in itself a real time value and is unlimited, uni-cognitional, and truth-provoking. In a subjective/objective reality, Earth wheels on its ears and enters the floodgates of the Infinite without having to wait for Spring. It is rooted in the now, and is Time itself.

Facial Characteristics

As Earth develops new sequences of formation, the notion of an empty face presents itself. In this valuation, Earth is not limited to any set of facial characteristics that might denote its likeness, except for the vibrational sequences that continue to establish a tonality of rapid advance in its own makeup. When we come to identify with the face of Earth as globobiotic—that is, infinitely structured in a global symbiosis with the awakening forms of life/being, the magnitude of constructive formation that can be achieved becomes enormous. The Earth can shake itself loose, undergo its own lift in the infrastucture, and precede the gateways in such a way that the face of God can reveal itself.

Through the reality of a parallel time portal, Earth can relive its own face, and the planet's most profound identity matrix becomes available. We are Earth, the internal apparatus that tugs Earth along, docking it squarely in its own port of paradise. In the globobiotic union with clock time and oceanic time, paradimensional reality is advanced. As this domain unfolds, sharing its history, its magnitude of variance, it is founded substantially on its true terms.

Earth is revivified; it enters through the bio-social gateway through which all of its relations with cognitive light are available; the outreaches of perception between space-time and clock time are unified. We can organically connect with our brothers and sisters in Creation without leaving home. All of our vantage points can be vested in one place, the globobiotic interlock of mind.

CONCLUSION

The interior perception of timeshift optimizes the state of consciousness to the point where the experiences of events, both inside and outside the body/mind field, are permanently joined. This experience gradually increases so that expansion into a state of uniform dimensionality catapults the participant into a state of free-moving, timeless awareness.

The psychological jumps necessary to create this state are based on stability, both physical and emotional. With increased experience of open-hearted assuredness, the depth impingements of the childself are cleared. Alienation or fear, brought about through the shift into non-timebound reality, rises to the surface for reintegration. An understanding of this process, a willingness to receive the support of others whose experience might be similar or at least akin, makes for a smooth transition.

Gradually one identifies with the underpinnings of events pictured in the outer world, increasing the awareness of significance. It is not that every small event is registered as equal to every other. However, there is a consistent feeling that the matter/time circumference is shrinking, that the center of being lies deep within an internal identification with Self, mirrored by a paradoxical luminosity. Objects of perception appear simultaneously distant while at the same time inexplicably intimate. There is a constant change in one's frame of reference.

In this luminosity lies a feeling of time leaning into itself, opening its essence like a great open-mouthed dragon, seeking treasure. The challenge is to adapt to this experience, to know that what one is forms the substance of everything. Yet, in this shift to uniformity, a fantastic quality of uniqueness can provoke intense feelings of anxiety and separateness. Resting in the power of this uniqueness brings on a gradual sense of calm. One can look at the ocean without becoming disoriented by the rhythms of its never-ending sameness.

In the cross-linking between these two paradoxical and

intense realizations—separateness and oceanic unity—the body itself begins to have an experience of shifting. It feels that it is both heavy and light, rigid and fluid, fully formed and without form. It goes through a shapeshifting process, during which the interior Self, as increasingly motionless onlooker, watches the body wriggle to adapt to its constantly changing reality.

During sleep, the viewer has less need to engage in the discernment necessary for practical activity. This safety can propel the timeshift experience to the point where the knower is standing directly in time-zero, looking onto the plane of past or future events with a sense of climbing into and beyond them, choosing a seat in an empty stadium. The seats, however, may all appear to be in a constant state of movement.

Where is the ego in all of this? The perceptions of personality, the notions of mind, its dislikes and likes, may not have completely fallen away. They are noted by the participant, in waking, dreaming or sleeping, but are viewed like advertisements on the mental screen; there is an automatic realization to ignore cues that do not help in the push towards expansion.

While awake or asleep, the participant can see the personality gripped by fear or other strong emotion; this may threaten a meltdown. This is where humor seems to be of great help. The attempt of the ego to rule non-timebound perception from its own singularity is absurd. One must learn to lovingly and humorously acknowledge and accept the personality's need to control, while choosing to hold to the inner heart of immediate progress.

The body created through consciousness realizes a merged view of existence, becoming gradually filled with an ever-expanding Light. This light, which many describe as Christic or God-like, is the body's essential nature. This bodylight begins to view itself as emanating from a central source within the delicate branches of the heart. The heart expands, back to front, side to side, through every vestibule, constantly reestablishing its boundaries until there are none. At times, the heart feels that it cannot withstand another minute of such constant reconstruction. It is like going home everyday to find one's intimate quarters completely rearranged. Where am I now? The identity must learn to find its moorings in its own breath, its own arrangement of "nowness," or it will continually try to keep the furniture from moving across the room.

From this state, an experience of an all-pervading God force, preeminent in nature, can begin to live permanently in the base of the heart. With alterations of time, and movement backwards or forwards in perception, the heart can remain steady, feeling God within and without.

As Earth shifts its own orbit, it can free-wield through the interludes of space/time with a vast dimensional berth. The participant can open to the elasticity of this condition, entering into a union with Earth that encourages continuity in the face of almost intolerable change. This agreement of the soul to support the changing Terran landscape, to make room for the recalibration of time on a planetary level, can alleviate greatly the temptation to escape through what we now call "death." Perhaps it will one day make physical death, as we know it, unnecessary.

Earth is promoting responsible disintegration of cherished notions of truth. In social and political life, this denouement is viewed as frightening, in some cases, even evil. The breakup of the traditional family is at the center of this perceived catastrophe. Although it is true that guardianship of the human spirit is essential for evolutionary unfoldment, this guardianship can come in a variety of forms. Extended family systems, based in spiritually aligned communal installations, are a safe and habitable form to work towards during this period of transition. Ultimately, consciousness itself must become the guardian; it must discern its own anchor in ethical and celestial perception.

God may literally be found in careful attention paid to inner truth. Young people particularly desire contact with this bigger Self which mediates the erratic behavior resulting from perceived lack of meaning. With courage gained, one can surrender to the mysterious, purely adventurous, yet miraculously conservative Christic force. God conserves universal strength through pointing out principles of conduct that lovingly continue to direct the seeker into a pattern of surrender; this happens much more easily once obscurations of fear or limitation are removed. Many people at this time are struggling against such freedom because they mistakenly see it as a form of anarchy. It may be viewed as a constant, circumspect value of rightness, felt by the individual and society as a prompting generated through depth realization of Self.

It appears now that fears of catastrophe—war, disease, or natural disaster—have risen to the surface of collective consciousness. Indeed, at times it seems that such events are more prevalent rather than less. When the family hound comes home from a day in the woods, he has to shake off his coat and clean his fur before settling onto the family hearth. So it is with our planet right now. Earth is in a state of self-grooming, and it is becoming clear that behavior that creates irrevocable harm will not be tolerated.

The window of change involves a dynamic, rapid movement into creative endeavors that will noiselessly heal the pangs of survival. Sharing, caring, opening oneself to the arms of the other without a sense of otherness. We are living in times of emergency, and what is emerging is a cooperative, non-combative repositioning for a journey into time.

God is demanding that we love each other so intensely, that when we look into each other's eyes, the desire to do something that would intentionally cause pain, simply becomes impossible. When one is standing naked in the field of time, stripped of armoring, and open to vistas of mathematical circumstance, there can be cause for both rejoicing and confusion. Climbing past this threshold of fear, consciousness can live in emptiness and be blanketed by warmth. Detachment is cool to the touch only until compassion sets fire to it.

The experience of dimensional change is an experience of reason voiced from the unreasonable. It cannot be mandated, nor can it be ignored. Free from any notion of fixed identity, one cannot cling to what is not there. Curiously, this state brings about a revivified sense of purpose; a sureness that the life one is living has meaning and is not to be lived without attentiveness to its rarity and sacredness. Consciousness can realize its causal relationship with matter; it experiences the land of no-time as fertile ground for creation. True freedom brings happiness when the desire to be unfree no longer remains. This happens gradually, but as our planet's sense of mission is unfolded, we may feel as if we have been placed in a slingshot; we will be stretched to the maximum.

The reminders we offer others that we are on a journey that even in death does not end, can be quickened with the commitment of love. Love that is constant and assured will help us enter interdimensional eternity. Braced by Self, centered in the universal willingness to assist and be assisted, time will stand still long enough for us to slip into the back of change. Far from being static, the future has many possibilities, each based on collective ripening of inner knowledge. To harvest the future, we must look carefully at the present, learn from it, and signal future outposts as to the insights we have gained. This is the means by which future time can be unified through a translucent present.

GLOSSARY

This glossary is meant to be used as a cue to the reader, viewed in context with the specialized usage in this book. The best way to understand these terms still rests with their being absorbed repetitively for their pure sound value.

arc—(see What Is an Arc?, p. 22) The doorway of consciousness through which time can travel. A mirror for the stream of data that creates the space-time portal to loop back over itself, causing a dimensional break into the free zone through which matter can enter.

bioreticulative—(see Geomorphic Resonance Fields, p. 178) Forming the "fiber optics" of the geomorphic resonance patterns. Creating the layering whereby the recrystallized formations can move freely into their newly mapped states.

breakfront(s)—(see Organization of Time and Space, p. 4) Spirals of energy/matter/time through which time can loop off to create its next qualifying interval. Units of space-time enclosure that allow time to seal up its matrices and spill them back into the space-time enclosure when necessary.

cellular memory—(see Who and What Earth Really Is, p. 159) The memory patterns of the cells, encoded with the time intervals necessary to recalibrate dimensional change. The means whereby time can teach the cells how to return to their original, less time-bound state.

coast mediums—(see Life Integers, p. 193) The interlays of matter/time that create the possibility of spin in the intervibrational webwork. The jumping-off point for time to lay its foundation in the pool of time variables.

crashfield identity—(see Living in the Continuum, p. 164) A feeling of discontinuity brought about through a feeling of attachment to planetary momentum from a fixed time perspective. The necessity of breaking from the traditional time parameters and falling into the relative from a free-fall, optimal vantage point.

early-time breakthrough—(see Interplanar Psychology, p. 58) The experience of time as conceived in earlier and earlier intervals so that the experience of all-time can be realized.

free-gap—(see Uniform Dimensionality, p. 56) A sense of relationship with an open-ended value of time in which the individual can freely map his/her position in the ocean of perception.

frequency response levels—(see Sacred Bonding with the Stars, p. 136) The parameters whereby consciousness restores itself to the field of matter through a step-down matrix of qualifying intervals. The frequencies that are generated to maintain and replicate these intervals in the space-time void.

geomorphic/geomorphological—(see Friendship Matrices, p. 156) Relating to the internal changes in pressure, motion, and volume in which the Earth is engaged. The process whereby the planet is shifting its form from time-bound relative, subject to fixed orbit and position, to a shapeshifting, internally governed reality.

globobiotic—(see Introduction, p. xxviii) Pertaining to the change in biological rhythms of the planet and their effects on world consciousness.

God structure—(see Life Seeds, p. 107) The mechanics of consciousness whereby the understanding of our interrelationship with God is imprinted in our physiology.

host factor—(see Matter as a Timed Event, p. 7) The condition of matter as a material ground for the unfoldment of space-time.

Host Intelligence—see Unlocking the Gates, p. 43) The manifestation of God through arrays of intelligence centered on Earth. That which feeds the visionary mantle for planetary evolution.

impedal velocity—(see What Is an Arc?, p. 23) The speed necessary to create the dynamic spin that opens time to constant change/motion in its involutionary spiral towards uniform dimensionality.

impedance variables, wave generated impedance—(see Assessing the Status of the Body, p. 125) Through the slowing down or impedance of signals entering the body/mind the physiology can catch up with itself and promote better grounding of the consciousness signaling system.

intercedance—(see Life Integers, p. 193) The ability of consciousness to move between and within two opposing channels of awareness without losing its internal flow.

interlitic—(see Upping the Periscope, p. 186) The interwoven vortices of light/color/sound that magnify themselves in our field and promote expansion. The field-dominance of such light patterns in our awareness.

life seeds—(see Life Seeds, p. 22) The internal entryway through which the impulses of life material are fed into the collective webwork for processing. The genetic strains of influence through which other civilizations have opened themselves to the beings of Earth and influenced our evolution.

locator point—(see Locator Points in Matter/Time Consciousness, p. 42) Inceptionary points through which matter/time can enter creating expanded time frames.

loop-locking—(see Interlocking Time and Space, p. 8) The interface created through uniform dimensional experience in which time can fold back over itself so that forwards and backwards are now one fluid motion in the time bank synthesis.

mathematical symbiosis—(see Time Codes, p. 11) The intimate process through which mathematical constructs are ingested by the field of time and recalibrated so as to promote reunification. The process whereby sentient beings digest mathematical material and derive psychological meaning.

matrilinear—(see Exploring the Field of Every-Mind, p. 34) An expression describing matter as a field-timed event through which possibilities in consciousness are arranged sequentially.

matrix, extended—(see Organization of Time and Space, p. 5) A system of data, thematically organized to express intervals of consciousness in fixed or random styles of functioning.

message units—(see Ingesting the Mathematics of Time, p. 18) Communications data bases that express possible future interlinks.

oceodynamic—(see The Language of God, p. 101) Pertaining to the ocean-like flow of telepathic signals that can replace spoken language.

parametric splits—(see The Value of No Time, No Space, p. 162) Openings in the time vault whereby Earth will shift its rotational spin and "split its seams," revealing its true nature.

psychosolar—(see Ingesting the Mathematics of Time, p. 16) Having to do with the influence of solar energy on the recalibration of the psyche.

pure consciousness, pure awareness—(see Organization of Time and Space, p. 4) The transcendental experience of expanded, infinite realization without an object. Wordless, indescribable Light.

queue up—(see Indicative Trends in the Field of Time, p. 14) Time preparing itself in a structured fashion to reveal its point of origination in the operative matrix.

radionically—(see Activity Functions, p. 184) Having to do with psychical energy manifesting itself through the form of expanding or counter-spinning radii.

response pools—(see Who and What Earth Really Is, p. 158) Collective possibilities of reply to changes in Earth's consciousness.

sequestral matter—(see Upping the Periscope, p. 186) Material that has been held back or protected to prevent exposure to higher dimensional frequencies until conditions are appropriate.

shapeshifting—(see Psychological Approaches to Meeting Time, p. 69) A dramatic change in form propelled by vibrational resurgence, due to the need for a more flexible, uniform, structure.

shield response—(see What Is an Arc?, p. 23) A fixed time response created as a balance to offset matrical interlays that might impinge on the temporal field. In relation to the Earth, the fixed time vantage points locked in place to temporarily limit the movement of consciousness from this plane.

slope of time—(see What Is an Arc?, p. 23) The literal slide of time from a fixed vantage point to an interlocking time-free zone. The shape of time as it creates this freefall motion in space-time.

spin, centripetal down spin—(see What Is an Arc?, p. 23) The involutionary spiral of time as it enters the field of resurgence.

time codes—(see The Organization of Time and Space, p. 6) The calibrations that time follows in its mathematical sequentiation of unfoldment. The change in these codes is the fundamental component behind dimensional shift.

time curves—(see Understanding Units of Mass, p. 19) As time spins, it naturally curves back on itself. These curves are speeded up, allowing matter to manifest as a more direct and energized experience of space-time.

time shifts—(see Matter as a Timed Event, p. 6) The redirection of time/space/motion from one category to another. These shifts create changes in our felt experience of reality.

transdimensional—(see The Personality and the Persona, p. 46) Experience that occurs in such a way as to cause the individual or collective to redirect the sensibility of reality. This can be experienced as a stripping down of the ego or persona that has been viewed as the core identity.

upscope—(see Upping the Periscope, p. 187) The sense of moving up in the evolutionary spiral towards a wider, more luminous vibrational context.

vibro-dimensional ribbing—(see Life Integers, p. 193) The striations in the fabric of time that allow us to experience its texture. This gives us the possibility of seeing how we can unfold time in such a way that new identities, new spirals in the space-time map, can reveal themselves to us, much like the weaving of new clothes.

void time—(see Living in the Continuum, p. 163) As we seek to meet the Self, we enter a condition whereby we gain a greater degree of internal space. This opens us to seeing new landscapes within ourselves as well as seeing the new Earth.

wraparound value—(see Holding Ourselves Together, p. 117) Consciousness wraps around itself like a cocoon, spun to protect and sustain vibrational gain. This cloaking leads to the development of our expression of physicality.